Emma Poole • Caroline Reynolds •

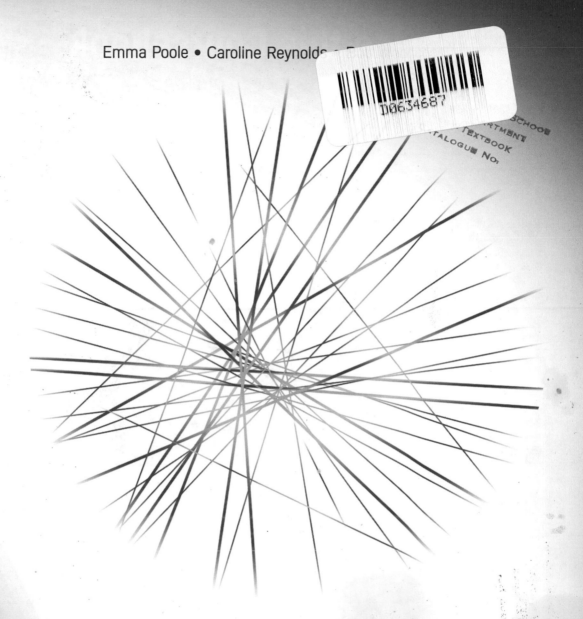

ESSENTIALS

KS3
Science Coursebook
covers all three years

How to Use this Coursebook

How to Use this Coursebook

A Note to the Teacher

This coursebook provides full coverage of the programme for study for Key Stage 3 science. It brings together all the relevant information in a single book to provide a flexible resource that can be used to support any model for delivering the curriculum.

The coursebook includes lots of questions and tasks. These provide opportunities for skills practice relevant to the Assessing Pupils' Progress (APP) assessment criteria and focuses, as well as helping to reinforce students' learning and improve their confidence.

The content is split into 36 topics, which are grouped together and colour-coded for ease of reference:

- Biology (green)
- Chemistry (red)
- Physics (blue).

Each topic consists of seven pages.

The first four pages of each topic provide clear, concise coverage of all the relevant content, incorporating Key Words boxes and Quick Tests.

The final three pages in each topic contain questions and exercises to reinforce students' understanding and provide skills practice:

- **Key Words Exercise** – to develop the students' scientific vocabulary.
- **Comprehension** – to ensure students can interpret information using scientific ideas.

- **Testing Understanding** – a literacy exercise, plus another exercise to help develop key skills.
- **Skills Practice** – allows students to practise and develop their investigative skills. These tasks can be used as the basis for practical sessions and discussions to provide opportunities for evaluation.

The answers to all the Quick Test and practice questions are included at the back of the coursebook.

This coursebook is supported by three workbooks for Years 7, 8 and 9, which feature levelled questions to support development and progression.

The workbooks provide further skills practice relevant to the APP assessment criteria and focuses, and help to consolidate students' learning.

To make cross-referencing between the coursebook and workbooks easy, details of the relevant workbook pages are given on the last page of practice questions in each topic of the coursebook.

Together, the coursebook and workbooks can be used to...

- help identify relative strengths and weaknesses for curricular target setting
- generate evidence of attainment as part of day-to-day assessment
- build evidence of student achievement for periodic, level-related assessment.

A Note to the Student

We're sure you'll enjoy using this coursebook. Follow these helpful hints to make the most of it:

- Try to learn what all the key words mean.

- Use the tick boxes on the contents page to track your progress: put a tick in the box next to each topic when you're confident you know it.

- Try to write your answers in good English, using correct spelling and good sentence construction. Read what you have written to be sure it makes sense.

- Think carefully when drawing graphs. Always make sure you have labelled your axes and plotted points accurately.

Contents

Contents

Chemistry Year 7

Chemistry Year 8

Chemistry Year 9

Contents

Biology

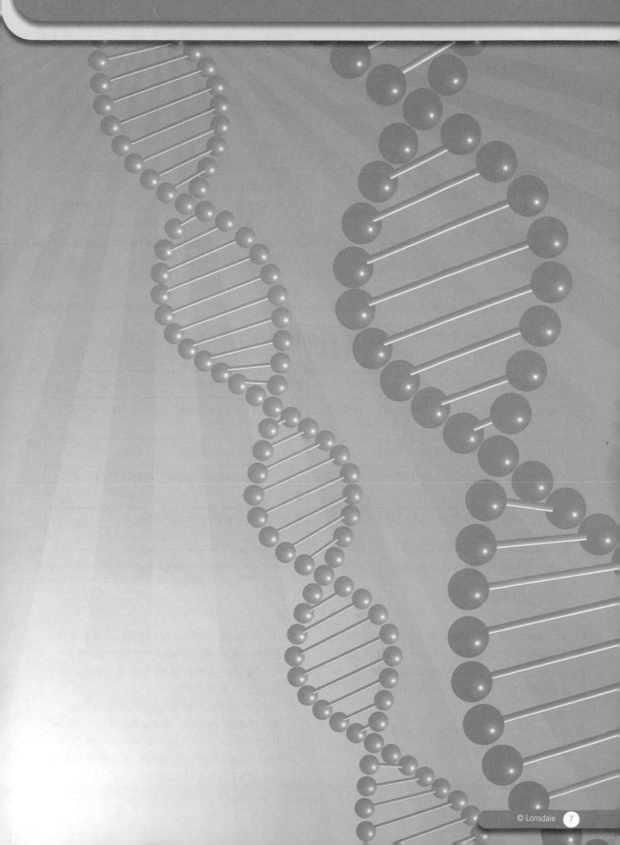

Cycle of Reproduction

Fertilisation

Fertilisation is the fusing (joining together) of the male nucleus with the female nucleus.

In animals, the sperm carries the male nucleus to the female egg. The point at which a new life begins is known as conception.

External fertilisation takes place outside of the bodies of the parents:
- Male fish release sperm and female fish release eggs as they swim alongside each other.
- Frogs cling together, and the male releases sperm and the female releases eggs at the same time.

Internal fertilisation takes place inside the body of the female:
- Sperm are released directly into the female's body where they must swim to the egg.
- Reptiles and birds produce their eggs after fertilisation.

Differences between External and Internal Fertilisation

External Fertilisation	Internal Fertilisation
Many eggs are produced – thousands or millions.	Fewer eggs are produced – between one and several hundred.
Often many eggs don't get fertilised.	Most eggs are fertilised.
Offspring are underdeveloped when they hatch.	Offspring are well developed when they hatch / are born.
Only a small percentage of young survive to maturity.	Most young survive to maturity.

Parental Care

In external fertilisation, many eggs are left alone and end up being eaten by other animals. Some animals protect the maturing eggs but abandon the young once they hatch.

In internal fertilisation, eggs are usually laid in some kind of nest, for example, in sand, or remain inside the mother.

- Birds incubate the eggs in the nest to keep them warm. After hatching, the young are fed by the parents until they are ready to leave the nest.
- Mammals feed the young with milk from the mother and the young are protected by the parents until they can survive on their own. This aftercare greatly increases the young's chance of survival.

Gametes

The sex cells are known as gametes. They are well adapted to their function.

The male gamete is the sperm. It has...
- a tail to enable it to swim
- structures in its cytoplasm to give it the energy it needs to swim
- a streamlined shape, due to having only a little cytoplasm, to help it swim easily
- chemicals in the tip of its head to help it break into the egg cell membrane
- in its nucleus, half of the genetic information that is needed to make the new individual.

The female gamete is the egg. It has...
- massive food reserves in its cytoplasm
- in its nucleus, half of the genetic information that is needed to make the new individual
- a layer of jelly around it to protect it.

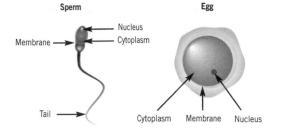

Cell Division

When a sperm enters an egg, its nucleus fuses with the egg cell's nucleus. This combines the inherited information from the mother and the father to form a new individual who has characteristics of both parents.

The newly fertilised egg then starts to divide.

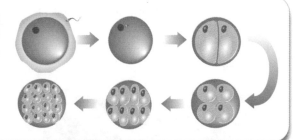

The Male and Female Reproductive Systems

- Sperm is continually produced in the testes.
- When sexually aroused, the penis becomes erect and is inserted into the vagina during intercourse.
- Sperm is then ejaculated (squeezed out) from the penis at the neck of the uterus (the cervix).
- The sperm swim from the cervix through the uterus and up into the oviducts.
- Here, they may meet and fuse with an egg cell if one has been released by an ovary up to three days prior to intercourse.
- Unfertilised egg cells last for a maximum of three days, but sperm may survive for longer.

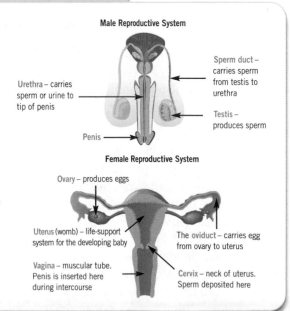

Cycle of Reproduction

Menstrual Cycle

A sexually mature female releases an egg from one of her ovaries approximately every 28 days.

Her uterus lining must prepare itself in case an egg is fertilised, to allow it to grow and develop. The menstrual cycle is as follows:

1. Uterus lining breaks down (i.e. period).
2. Repair of the uterus wall. Uterus lining gradually thickens.
3. Egg is released.
4. Lining stays thick waiting for fertilised egg.
5. If there is no fertilised egg the cycle restarts.

Fertilisation Cycle

Fertilisation takes place in the oviduct. This allows the egg time to divide and develop so that it's ready to grow into the uterus wall. So, sperm must be in the oviduct at the same time as the egg.

Sexual intercourse between days 12 and 15 of the menstrual cycle could result in the following:

1. Ovulation – an egg is released from an ovary. (Intercourse – sperm is ejaculated into the cervix.)
2. Fertilisation – the egg and sperm fuse together.
3. Cell division – division of fertilised egg to form a ball of cells that becomes an embryo.
4. Implantation – embryo implants into the spongy lining of the uterus.

Development of the Foetus

After about nine to ten weeks, the limbs become visible and the embryo begins to look like a human baby, so it is now called a foetus.

During development and growth, the umbilical cord and amnion play important roles:

- The umbilical cord carries the foetal blood very close to the mother's blood.
  - Food, oxygen and water pass from the mother's blood into the foetal blood
  - Carbon dioxide and waste pass from the foetal blood into the mother's blood.
  - The two bloods do not mix together.

- The umbilical cord acts as a barrier to some harmful substances. But other substances, for example, heroin, alcohol, substances in tobacco smoke, and viruses such as Rubella (German measles) and HIV, can pass between the bloods and harm the foetus.
- The foetus is supported and protected from minor bumps by amniotic fluid (a watery liquid) which is surrounded by the amnion (a membrane). This breaks just before birth, releasing the fluid. This is known as 'the waters breaking'.

Birth

During birth...
- the uterus muscles contract, pushing out the baby, followed by the **placenta** (the afterbirth)
- the cervix muscles relax to allow the baby to pass down the vagina.

Once the baby's born...
- its lungs fill with air
- it suckles to obtain milk.

Breast milk is rich in fat and protein, which helps the baby to grow. It also carries some protection from infections.

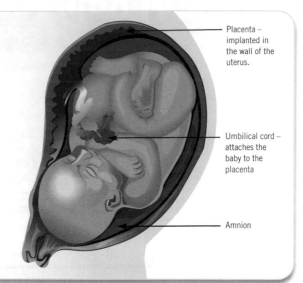

Placenta – implanted in the wall of the uterus.

Umbilical cord – attaches the baby to the placenta

Amnion

Adolescence

After birth, cells divide and grow at a steady rate, but there are some remarkably rapid spurts at times.

The child grows, and passes into adolescence before becoming an adult.

During adolescence, a human's reproductive organs become mature and, physically, the person changes from a boy or girl into an adult.

The physical changes that take place during adolescence are listed in this table.

Girls	Boys
Ovaries start to release eggs	Testes start to produce sperm
Breasts start to develop	Muscles and penis grow bigger
Hips get broader	Voice becomes deeper
Pubic and underarm hair grows	Pubic, facial and underarm hair grows

Quick Test

1. What is fertilisation?
2. Give one difference between internal and external fertilisation.
3. Give one way in which a sperm is well adapted to its function.
4. What is the function of the amnion during pregnancy?
5. Give one way in which a girl's body changes during adolescence.

KEY WORDS
Make sure you understand these words before moving on!
- Amnion
- Cervix
- Conception
- Embryo
- Fertilisation
- Foetus
- Implantation
- Menstrual cycle
- Ovary
- Oviduct
- Ovulation
- Penis
- Placenta
- Sperm duct
- Testis
- Umbilical cord
- Urethra
- Uterus
- Vagina

Cycle of Reproduction

Key Words Exercise

Match each key word with its meaning.

Embryo		The series of changes that occur in the female reproductive system each month
Fertilisation		A muscular organ, also known as the womb, where the baby develops
Foetus		The developing egg before it's recognisable as a human baby
Menstrual cycle		Carries sperm and urine to the tip of the penis
Ovulation		The release of an egg from an ovary
Placenta		The fusing of the egg nucleus with the sperm nucleus
Urethra		The developing baby once it has limbs and is obviously human
Uterus		Where substances are exchanged between the mother's blood and foetal blood

Comprehension

Read the passage about multiple births and then answer the following questions.

Identical twins occur when the ball of cells that is produced by a fertilised egg splits into two separate groups of cells. The two groups of cells implant into the uterus wall and develop as two separate, but genetically identical, individuals. If the two groups of cells fail to completely separate (which is very rare), the twin offspring may be joined at some part of the body. They are known as conjoined twins.

A woman can sometimes release more than one egg at a time. If two or more eggs are fertilised by separate sperm and then implant, the result is non-identical twins, triplets, etc. They will not necessarily be the same sex.

1. How many sperm are needed to produce identical twins? Explain your answer.

2. How many eggs are needed to produce non-identical twins? Explain your answer.

3. If two eggs are released at the same time, but only one of them is fertilised, what would happen?

4. If two eggs are released at the same time and both of them are fertilised but, before implantation, one of them divides into two, what would happen?

5. What problems might arise from a mother having twins, triplets, quadruplets (or more) developing in her womb?

Testing Understanding

1 **Fill in the missing words to complete the sentences about sexual reproduction in humans.**

a) Approximately every 28 days, one of the female's _____ releases an egg. This is called _____.

b) The egg is then moved along an _____. If the egg meets a sperm here, the sperm may break through the egg's membrane and the two nuclei will fuse together. This is called _____.

c) The fertilised egg divides rapidly and passes into the _____, which has formed a thick, spongy wall, so that the egg can _____ into it. For the first nine to ten weeks it's called an _____. When it has limbs it's called a _____.

d) The developing baby will stay here for about 40 weeks in total before it's born. After which, the baby will feed on _____ from its mother's breasts.

e) If the egg is not fertilised, the thickened wall will break down and the _____ cycle will begin again.

2 **The diagrams show specialised sex cells.**

Write the name of the structure that performs each of the following functions.

a) Contains food so the embryo can grow

b) Enables sperm to swim

c) Contains enzymes to break through the egg membrane

d) Contains half the genetic information (from the father)

e) Contains half the genetic information (from the mother)

f) Protects the egg

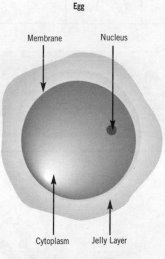

Egg

Membrane   Nucleus

Cytoplasm   Jelly Layer

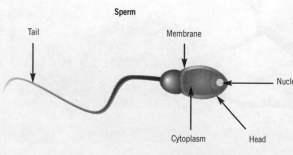

Sperm

Tail   Membrane

Nucleus

Cytoplasm   Head

Cycle of Reproduction

Skills Practice

Mairead and Khan wanted to investigate growth in the locusts that were kept in their classroom.

They thought that temperature might affect the locusts' growth rate.

To test this, they put food and shelter into five tanks and put five tiny locusts into each tank. Each tank had a thermostatic control.

The temperatures were set at 2°C, 6°C, 10°C, 14°C and 20°C.

Their results were as follows:

1. Name one factor that must be controlled (kept the same) to make this a fair test.
2. Copy the axes below and plot the results onto them. You will need to plot a new line for each temperature.
3. What conclusions can you draw from the results?
4. What further investigations could you make based on your conclusions?
5. How reliable are the findings of this investigation? Explain your answer.

Temp.	Average Length of Locust (cm)				
	2°C	6°C	10°C	14°C	20°C
Day 1	2.2	2.4	2	2.4	2.4
Day 5	2.4	3.6	3.6	4.4	5.6
Day 10	2.8	4.2	5.6	7.6	9

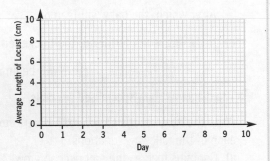

Organisation of Life

Plant Cells

Living things are called organisms. They're made up of cells. All plant cells have the following features:

- Chloroplasts for photosynthesis (except for root hair cells, which aren't exposed to light).
- Jelly-like cytoplasm.
- Thick outer cell wall made of cellulose.
- Cell membrane to control what enters and leaves the cell.
- Watery, central vacuole.
- Nucleus to control what the cell does.

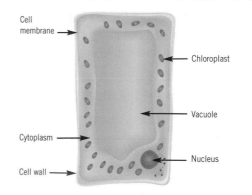

Animal Cells

A typical animal cell is simpler, and has only three features:

- Nucleus to control what the cell does.
- Cell membrane to control what enters and leaves the cell.
- Jelly-like cytoplasm.

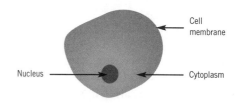

Comparing Animal and Plant Cells

Plant and animal cells are very similar, but they're adapted to carry out different functions, which is why they have different features.

Feature	Description	In Plant Cells?	In Animal Cells?
Nucleus	Controls all the cell reactions	Yes	Yes
Cytoplasm	Where chemical reactions happen	Yes	Yes
Cell membrane	Controls what enters and leaves the cell	Yes	Yes
Cell wall	Made of tough cellulose to give the cell a rigid shape	Yes	No
Vacuole	Large space containing cell sap (mainly water)	Almost always	No
Chloroplast	The structures where photosynthesis happens	Only those exposed to light	No

Organisation of Life

Cell Specialisation

Cells are specialised to carry out a particular job. A specialised cell may be a specific shape and may make different chemicals in its cytoplasm.

Some examples of specialised cells are shown below.

- Some living things are made up of just one type of cell.
- A daffodil has around 80 different types of cell.
- A human body has around 200 different types of cell.

Epithelial cell – a lining that covers large areas, e.g. skin. 	Root hair cell – the cell's shape helps to increase the surface area to make it easier to absorb water. 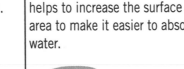	Neurone (nerve cell) – a long cell that transmits messages from one place to another. 
Palisade cell – packed with chloroplasts, especially in the upper half, to catch light. Found in the upper surface of a leaf. 	Sperm cell – contains the male nucleus and has a tail so that it can swim. 	Egg cell (ovum) – contains the female nucleus and food materials for the developing embryo.
Red blood cell – has no nucleus to enable it to carry more oxygen around the human body. 	White blood cell – can change its shape to engulf bacteria and then digest them. 	Ciliated cell – found in the throat and nose. The cell produces mucus which the cilia waft towards the mouth and nose. Also found in the oviduct to move the egg along. 

Tissues to Organs

Lots of similar types of cell are found in groups that work together to carry out a specific job. They are known as a **tissue**, for example...

- muscle tissue
- palisade tissue
- skin tissue.

When two or more tissues work together to carry out a particular function, they're known as an **organ**, for example...

- stomach
- leaf.

Organs are arranged together to form a system, for example...

- digestive system
- circulatory system
- reproductive system.

The systems work together so the body can do all of the jobs it needs to do in order to stay alive. The whole being is known as an **organism**.

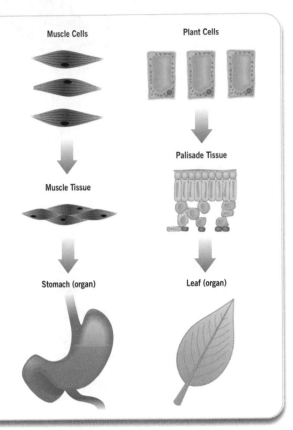

Muscle Cells

Plant Cells

Muscle Tissue

Palisade Tissue

Stomach (organ)

Leaf (organ)

The Digestive System

The digestive system contains several different organs. Each organ has a different job, described in the table below.

All the organs work together to process food through the body to give you the energy and resources you need in order to grow and repair yourself.

Organ	Main Functions
Mouth	Teeth begin to break down food and saliva starts to break down starch
Oesophagus (foodpipe)	Pushes food from the mouth to the stomach
Stomach	Begins to break down proteins and acid kills bacteria
Small intestine	All food is digested here and is absorbed into the bloodstream
Large intestine	Absorbs water and allows waste to be removed from the body

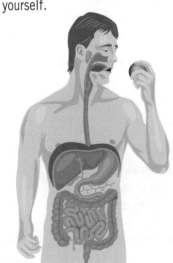

Organisation of Life

Animal Cell Division

Animal cells don't live for the entire life of an organism, so they're replaced in this way:
1. The nucleus divides into two.
2. The cytoplasm then divides, forming two new 'daughter' cells, each identical to the original.
3. Each cell then grows to the same size as the original cell.

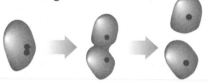

Plant Cell Division

Plant cells divide in a similar way to animal cells, but after division the vacuole forms and the cell further from the tip of the plant grows.

The cell divides again, and the process continues. In this way, the plant grows at the tip, while the older cells become specialised to form new tissues.

Fertilisation in Flowering Plants

Pollination is the moving of pollen grains from the anthers of one flower to the stigma of another.

Fertilisation is the fusing of the male pollen grain nucleus with the female ovule nucleus. The pollen grain grows a pollen tube through the style and ovary to carry the nucleus down.

The pollen grain and ovule cells are specialised to allow genetic information to be passed on to the next generation from both parent plants.

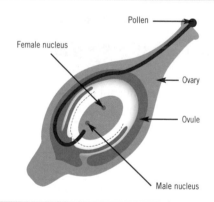

Pollen

Female nucleus

Ovary

Ovule

Male nucleus

Quick Test

1. Name three features that both plant and animal cells have.
2. a) Draw a neurone (nerve cell).
   b) How is a neurone specialised to carry out its job?
3. What is an organ made up of?
4. What is pollination?

KEY WORDS
Make sure you understand these words before moving on!
- Cell
- Cell membrane
- Cell wall
- Chloroplast
- Cytoplasm
- Nucleus
- Organ
- Organism
- Photosynthesis
- Pollen tube
- Pollination
- Specialise
- System
- Tissue
- Vacuole

Key Words Exercise

Match each key word with its meaning.

Cell	The 'building block' of life
Cell membrane	A rigid structure made of cellulose around plant cells
Cell wall	Surrounds the cell and controls what can enter or leave it
Chloroplast	Controls all the cell reactions and contains the genetic information
Cytoplasm	The process that green plants use to make sugars (food)
Nucleus	Green structures found in plant cells that absorb light for photosynthesis
Organ	Grows down the style and carries the male nucleus to the plant ovule
Photosynthesis	A jelly-like substance where chemical reactions take place
Pollen tube	A group of similar cells that carry out a particular function
Pollination	A group of tissues that carry out a particular function
System	A group of organs that carry out a particular function
Tissue	The transfer of pollen from an anther to a stigma

Comprehension

Read the passage about cells and then answer the following questions.

1. Why weren't cells recognised before the 17th Century?

2. When were the first living cells observed?

3. Why didn't the observer of these cells suggest that they were the unit of life?

4. What was the single most important event in the discovery of the cell?

5. How did people explain the presence of cells before cell theory became accepted?

6. What did Virchow mean when he said 'all cells come from cells'?

7. Make a list showing the order in which these important discoveries were made. Make sure you include the dates.

The microscope, which was invented in the 17th Century, allowed Robert Hooke and others to observe structures as small as cells. Hooke used the word 'cell' to describe the dead cells he observed in 1665. Neither he nor anyone else at the time realised that the 'empty' cells he saw had once contained the basic units of life.

Even when single-celled animal life was observed in 1673, it was thought to have 'just formed' because no-one had seen cell division.

Improved microscopes in the 1830s allowed the nucleus to be discovered; cytoplasm was discovered shortly afterwards.

In 1839, Schwann and Schlieden suggested that organisms are made from cells, and that organisms can be either single-celled or multi-celled. They recognised that the membrane, nucleus and cytoplasm are common to all cells.

In 1855, Virchow said that 'all cells come from cells' and by 1890, all life was thought to be made up of cells.

Organisation of Life

1 **Fill in the missing words to complete the sentences about plant and animal cells.**

a) Animal and plant cells have similar structures. They both have a _____

_____, which allows some substances in and out of the cell. The

_____ controls everything the cell does. All the reactions take place in

the _____.

b) A _____ cell is adapted to have special features that allow it to do its

job. The cells are grouped into _____, which are grouped into organs to

form a _____. An entire living being is called an _____.

c) Cells grow and replace themselves by dividing, which involves the splitting of the

_____ first and then the _____ to form two

'_____' cells.

d) Pollination, often by wind or insect, occurs when _____ grain cells are

transferred from the _____ of one flower to the _____

of another. It then grows a _____ _____ down the

_____ where its nucleus fuses with the _____ nucleus.

This is called _____.

2 **Look at these diagrams of cells, then answer the questions that follow.**

A

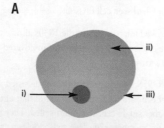

B

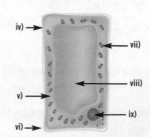

a) Write the names of the parts labelled.
b) **i)** Which one is an animal cell?
   ii) Explain your reasons.

Hannah and Finn decided to see what concentration of sugar solution was best to grow pollen tubes from lily pollen.

They put lily pollen grains in six different concentrations of sugar solution and looked at them using a microscope to see how many pollen grains had grown a pollen tube.

Hannah used 25 grains each time, but Finn chose to use only 2 grains. Here are their results:

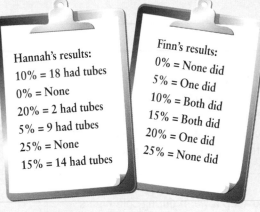

Hannah's results:
10% = 18 had tubes
0% = None
20% = 2 had tubes
5% = 9 had tubes
25% = None
15% = 14 had tubes

Finn's results:
0% = None did
5% = One did
10% = Both did
15% = Both did
20% = One did
25% = None did

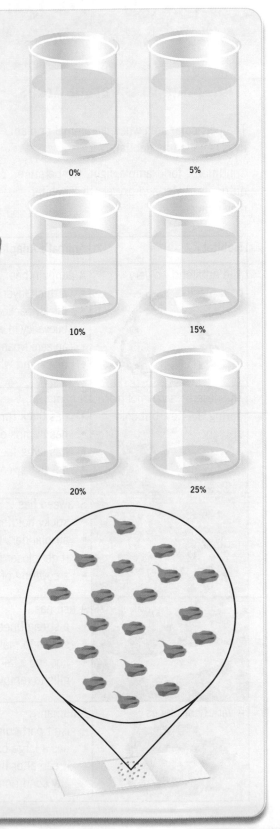

0%   5%

10%   15%

20%   25%

1 **a)** Who had the better chance of obtaining a fair result, Hannah or Finn?
   b) Give reasons for your answer to part a).

2 What is the dependent variable in this investigation?

3 What is the independent variable in this investigation?

4 **a)** Choose the best set of results, and put them in the correct order in a table. Make sure you label the columns in the table.
   b) Plot the results onto graph paper. Make sure you label the axes on the graph.

5 What conclusion can you make about the best sugar concentration for pollen tube germination in lily pollen, based upon this evidence?

6 What variables should be controlled in this investigation?

Interaction in the Environment

Adapting to Different Areas

The particular area where an animal or plant lives is called its **habitat**.

Conditions, for example, light, temperature, nutrients, water and oxygen, in a habitat may change.

Living things must be able to survive in the changing conditions in their habitat. Some examples of habitats and the ways in which animals have adapted to the conditions are shown in this table:

Habitat	Animal's Adaptations
The Antarctic – a very cold place	A penguin has... • a thick layer of fat for insulation and food storage • thickly packed feathers to trap air for insulation and to give it buoyancy in water • an aerodynamic (dart-shaped) body and webbed feet for swimming • black and white colouring for camouflage.
The desert – a very hot place during the day, but cold at night	A camel... • has thick fur to keep it warm during cold nights • has humps of fat for food storage and to release water • has large feet to spread out its weight so it can walk on sand • doesn't sweat much, to save water.
The seashore – a changing place (wet and dry)	Seaweed has... • sticky holdfast (like a root) to hold on to rocks • air bladders ('bubbles') to give it buoyancy in water and to help it get closer to the sunlight so it can photosynthesise • a covering of slime to reduce evaporation.
The sea – a very wet place	A fish has... • a streamlined shape to help it swim easily through water • scales to make it smooth to reduce friction • fins and a tail to allow it to steer through the water and propel itself • gills to remove oxygen from the water for respiration.
A variety of places	A human... • isn't particularly well adapted for any special environment • has a large brain and the ability to use tools and materials so it can alter its environment, allowing it to live in almost any conditions.

Adapting to Food

Animals are adapted to their food source:

- Eagles have a very sharp, hooked beak to tear flesh from their prey.
- Lions have very sharp, long canine teeth to tear flesh from their prey.
- Parrots have a strong, rounded beak to break open nuts and seeds.
- Sheep have flattened teeth to grind the plants they eat.

Adapting to Changes over a Day

The physical environment around your school varies dramatically over any 24-hour period. Light intensity, temperature, humidity and noise level can all change, as the top graphs show.

These changes could be responsible for behavioural changes in the animal population and could result in different animals visiting the school at different times, as shown in the lower graph.

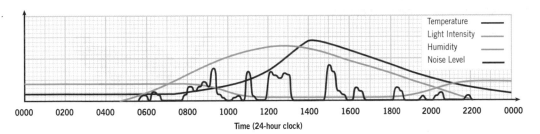

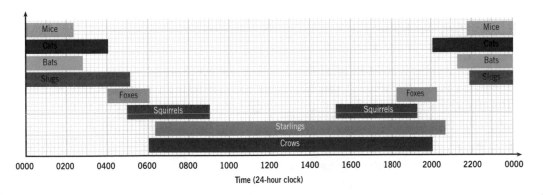

Adapting to Changes over a Year

The changes between the seasons may occur at a much slower rate than the changes during a day, but they affect the living things in a habitat much more. Different organisms use different techniques to survive these changes:

- Geese migrate (move) to warmer climates.
- The arctic hare grows a thicker winter coat for insulation.
- The hedgehog hibernates (sleeps through the winter) after building up fat reserves.
- The daffodil disappears underground and lives off the food store in its bulb.
- The oak tree sheds all its leaves in autumn to reduce water loss. It can survive without using much energy because there isn't much sunlight.

Interaction in the Environment

Food Chains

Food chains show what is eating what in any area.

All food chains begin with green plants (**producers**), which can use the Sun's light energy to make their own food by **photosynthesis**.

The animals in a food chain are all known as **consumers** because they eat (consume) other organisms in order to get their energy:

- Animals that eat plants are called **herbivores**.
- Animals that eat animals are called **carnivores**.
- Animals that eat both plants and animals, are called **omnivores**.

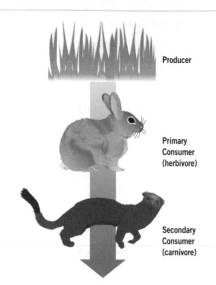

Producer

Primary Consumer (herbivore)

Secondary Consumer (carnivore)

Predators and Prey

- An animal that hunts and eats another animal is called a **predator**, e.g. osprey.
- The animal that's hunted and eaten is called the **prey**, e.g. fish.

Biological control is an **organic** method of pest control. It uses the relationship between organisms to control some pests, as in this example:

1. Greenfly (aphids) suck the sap of tomato plants, using up the plant's energy, which reduces the standard and quality of the crop.
2. Grower introduces ladybird larvae, which eat aphids when they become adults.
3. Number of aphids is reduced, meaning the tomato plants are of a higher quality.

This method doesn't kill all the aphids, but the reduced number of aphids cause less damage, which means chemicals don't need to be used.

The number of predators there are is very closely linked to the number of prey available:

1. As the number of prey increases, there is more food, so the number of predators increases.
2. As the number of predators increases, more prey are eaten, so the number of prey decreases.
3. As the number of prey decreases, there is less food, so the number of predators decreases and so on...

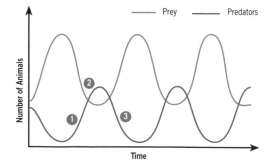

Food Webs

All the different food chains in a particular habitat can be connected to make a **food web**:

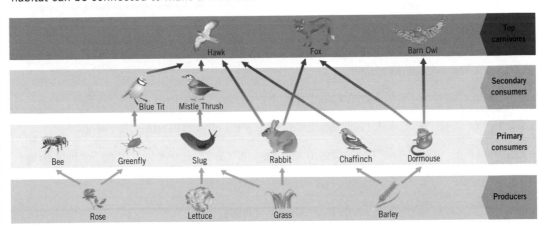

A food web shows many food chains interlinking with one another. It can be used to predict how other organisms will be affected if one species begins to die out.

For example, this is what would happen if the number of slugs were to be drastically reduced:

1. There would be more grass for the rabbits to eat. The number of rabbits would increase, giving hawks and foxes more to eat.
2. The number of mistle thrushes would rapidly decrease because they would have less food, so hawks may eat more rabbits, bluetits and chaffinches instead.
3. The number of greenfly may increase, which would reduce the numbers of roses, bees etc.

All the organisms in a habitat are **interdependent**. This means that one small change may have many consequences.

You can use the webs to make predictions about various effects on the habitat, which is a very useful tool when looking at conservation of a habitat.

For example, if another organism, e.g. a ladybird, is introduced, this would have an effect on the numbers of bluetits, hawks, mistle thrushes, etc.

So, before biological control can be undertaken, the overall effects should be taken into account to try to reduce any major impact on the habitat.

Quick Test

1. What is the name of the area in which a plant or animal lives?
2. Give two examples of how the penguin has adapted to its environment.
3. Give one method that animals use to survive the winter.
4. What are producers?
5. Give an example of a carnivore.
6. What is biological control?

KEY WORDS
Make sure you understand these words before moving on!

- Biological control
- Carnivore
- Consumer
- Food chain
- Food web
- Habitat
- Herbivore
- Interdependent
- Omnivore
- Organic
- Photosynthesis
- Predator
- Prey
- Producer

Interaction in the Environment

Key Words Exercise

Match each key word with its meaning.

Biological control	A green plant able to produce its own food
Carnivore	All animals, because they eat other organisms for energy
Consumer	The process by which plants make their own food using Sunlight energy
Food chain	Animals that eat only plants
Habitat	Animals that eat only animals
Herbivore	Animals that eat both plants and animals
Interdependent	An animal that hunts and eats other animals
Omnivore	An animal that's hunted and eaten by another animal
Organic	The particular area in which an organism lives
Photosynthesis	Method of using an organism to reduce the number of pests
Predator	Dependent on each other to stay alive in a habitat
Prey	A process that doesn't use artificial chemicals
Producer	A way of representing the energy transfer between organisms

Comprehension

Read the passage about predators and prey and then answer the following questions.

1 Are moose predators or prey?

2 Are wolves predators or prey?

3 Are moose herbivores or carnivores?

4 Are wolves herbivores or carnivores?

5 Suggest a reason for the low of 14 wolves in 1982.

6 Why do the numbers of moose and wolves change so much?

Isle Royale is an isolated island on Lake Superior in Canada. From 1900 to 1949, only moose lived on the island, feeding on the plants. But, during the winter of 1949, the lake iced over, enabling a few wolves to go over to the island. Wolves hunt and eat only moose.

The populations of both animals fluctuate each year. In the last 50 years, the lowest number of moose recorded was 540 in 2005 and the highest was 2440 in 1995. The lowest number of wolves was 14 in 1982 (after a dog had visited the island in 1980, which is usually not allowed) and the highest was 50 in 1980.

Scientists thought that the numbers would level at about 1500 moose and 25 wolves, but this hasn't happened.

Testing Understanding

1 **Fill in the missing words to complete the sentences about survival.**

a) A _____ is the place where plants and animals live.

b) Organisms are _____ to survive daily and seasonal changes in their
environmental conditions. These may be changes in _____ or in the
amount of _____ available.

c) Some animals _____ to warmer climates in winter, others go into a deep
sleep, called _____, to survive.

d) Animals are adapted to their type of _____. If they are carnivores they
have _____ teeth or beaks. If they eat plants they are called
_____ and have _____ teeth for grinding or a round beak.

2 **Read the information and then answer the questions that follow.**

The feeding relationships in an area of
woodland are as follows:

- Oak trees are eaten by leaf roller moths
  and winter moths.
- Hawthorn bushes are eaten by winter
  moths, voles, mice and squirrels.
- Leaf roller moths are eaten by mice
  and voles.
- Winter moths are eaten by mice, voles,
  rove beetles, great tits and shrews.
- Mice are eaten by weasels, owls
  and hawks.
- Voles are eaten by weasels, owls
  and hawks.
- Great tits are eaten by weasels and hawks.
- Rove beetles are eaten by shrews.
- Shrews are eaten by owls and weasels.

a) Construct a food web of this information.
b) i) Name a predator.
   ii) How do you know this is a predator?
c) Name an animal that feeds at two
   different levels.
d) What is the longest food chain in
   this diagram?
e) What might be the possible effects if all the
   squirrels died?
f) What is the source of energy that this whole
   food web relies upon?

Interaction in the Environment

Several students carried out an investigation into the sort of conditions that woodlice prefer. They used a 'choice chamber' to perform this task. The chamber is divided into light and dark halves, and wet and dry halves. This provides four choices for the woodlice:

- wet and dark
- wet and light
- dry and dark
- dry and light.

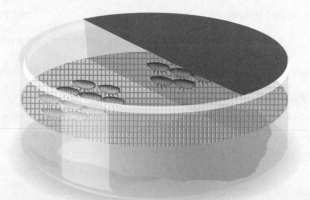

The students recorded where the woodlice were every two minutes. Their results are shown in this table:

Time (min)	0	2	4	6	8	10
Wet and dark	5	11	16	18	19	20
Wet and light	5	3	1	0	0	0
Dry and dark	5	4	3	2	1	0
Dry and light	5	2	0	0	0	0

1️⃣ What factors must have been kept the same in order to make this a fair test?

2️⃣ Plot the results on graph paper, using a different colour for each set of conditions.

3️⃣ Which conditions do woodlice like least?

4️⃣ What conclusions can you draw from this investigation?

5️⃣ a) If the experiment had been done with only four woodlice, would the results have been as useful?
   b) Explain your answer.

Sorting Differences

Variation

Variation means differences between individuals within the same species.

Dogs are all the same species as they have similar characteristics and can breed together to produce fertile offspring (i.e. offspring that are able to breed together).

A horse and a donkey may look similar, but they are different species. If they breed together they can produce a mule, which is infertile (unable to produce offspring).

Male Donkey

Female Horse

Mule

Measuring Variation

Measuring the hand spans of all the people in your class is a good way to see variation within a species.

You can record your results in a table (or spreadsheet) like this one:

Name	Hand Span (cm)
Carl	13.6
Alicia	12.2

Then sort the results into groups:

Hand Span (cm)	Number of Pupils
11.0–11.9	2
12.0–12.9	3
13.0–13.9	6
14.0–14.9	8
15.0–15.9	4
16.0–16.9	1

You can then present the data as a graph:

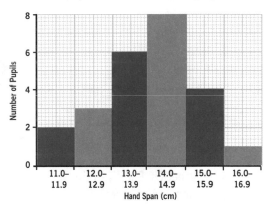

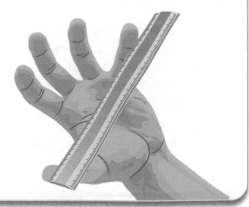

Sorting Differences

Causes of Variation

There are two reasons for differences between individuals. Each individual…
- inherits different characteristics from their parents, known as inherited differences
- lives in different conditions and has different experiences, known as environmental differences.

Genetic Factors

Children are usually similar to their parents, but they have a mixture of their parents' genetic information, which is why children don't have the exact same features as their parents.

Here are some examples of inherited features:
- nose shape
- ear shape
- eye colour
- hair colour.

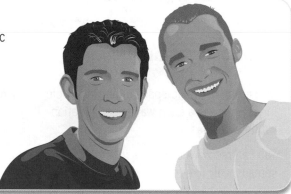

Environmental Factors

The natural shape and colour of your features are decided by genetics. But if you dye your hair or break your nose, your features will also have been affected by the environment.

Factors such as the ability to skate, or getting a scar are totally determined by the environment. Identical twins will start to differ as they get older – differences will occur due to differences in diet, lifestyle and so on.

Plants are affected greatly by their environment:
- Without sunlight or nutrients they don't grow very much.
- Without water they wilt.

Sorting into Groups

There are many different types of living things. Sorting them into groups helps you to identify them.

Organisms can be grouped in many ways. For example, they could be grouped by...
- the number of legs they have
- whether they have wings
- where they live
- any observable characteristics.

Classifying Living Things

Carolus Linnaeus, a Swedish naturalist, was the first person to develop a way of classifying living things into groups. He gave every living thing he identified a Latin name. He published his work in 1737 and his classification system is still used today.

The system is based on simple observations, such as the following:
- How many legs does it have?
- Does it have body segments?
- Does it have a backbone?

As well as helping us to identify things more easily, classification helps scientists to study the relationships between various organisms and to understand how they 'fit' and interact in the world.

More detailed information, for example, genetic information, is also collected and used to classify living things.

As more details are found out, some organisms may get reclassified.

Sorting Differences

Classification

Different species that are similar to each other can be grouped together.

For example, a monkey and a hamster are different from each other, but they have several features in common, for example...

- hair-covered bodies
- warm-blooded
- give birth to live young
- suckle their offspring with milk from the mother.

So, these animals are grouped together with other animals that also have the same features in common (including humans). They are called mammals.

There are several different groups, for example, amphibians, birds, fish, mammals, reptiles. The characteristics of each group in the animal kingdom are described in the table opposite.

The groups are then grouped together by their similarities to make larger, more general groups, such as vertebrates (animals with backbones) and invertebrates (animals without backbones).

All living things can be classified into one of five larger groups called kingdoms.

Animal Group	Characteristics
Amphibians	• Skin doesn't have scales, hair, or feathers • Moist skin to aid breathing • No claws on their toes
Birds	• Skin is covered with feathers • Lightweight (hollow) bones • Wings and a beak • Lay eggs
Fish	• Live in water • Skin is covered with scales • Fins for swimming • Gills for breathing • Lay eggs in water
Mammals	• Warm-blooded • Breathe air through lungs • Hair / fur on their bodies • Give birth to live young • Suckle their young
Reptiles	• Skin is dry and scaly (no hair or feathers) • Claws on toes • Lay eggs on land

Quick Test

1. Why can't horses and donkeys be classified as the same species?
2. What are the two reasons for differences between individuals of the same species?
3. What did Carolus Linnaeus do?
4. Give two examples of features that a hamster and a monkey have in common.

KEY WORDS
Make sure you understand these words before moving on!

- Amphibian
- Bird
- Classification
- Fish
- Infertile
- Invertebrate
- Kingdom
- Mammal
- Multi-cellular
- Reptile
- Species
- Variation
- Vertebrate

Key Words Exercise

Match each key word with its meaning.

Amphibian	Differences between individuals of the same species
Bird	Grouping living things using features they have in common
Classification	An animal that has a backbone
Fish	An animal that doesn't have a backbone
Infertile	A hair-covered animal that suckles its young
Invertebrate	An animal that has feathers, wings and a beak
Kingdom	An animal that has fins and gills
Mammal	An animal that has moist skin and lungs; it returns to water to breed
Multi-cellular	An animal that has lungs and dry, scaly skin
Reptile	A group of similar organisms able to produce fertile young
Species	An organism unable to produce young
Variation	The first grouping (of five) in the classification of living things
Vertebrate	A body made up of many cells working together

Comprehension

Read the passage about classification and then answer the following questions.

1. Why did Linnaeus create a classification system?

2. Why shouldn't you use size to classify a group?

3. Humans are classified as *Homo sapiens*. Which of the two names is the genus name?

4. As the groups of organisms get smaller, would you expect the organisms in each group to be more alike or more different?

5. What are keys used for?

Today's system of classification of all living things was devised by Linnaeus in 1751. It identifies similarities and groups together the organisms that share them.

Care was taken not to use criteria that showed possible difference due to the environment, for example, height.

Each group is then re-sorted using different criteria. Eventually, as each group becomes smaller, individual species can be identified.

Each species is identified firstly by its genus name, which it shares with other very similar organisms and secondly by a unique species name. For example, a domestic cat is *Felis cattus* and a tiger is *Felis tigris*.

In this way, a key to living things can be written to help people to identify animals or plants they find.

Sorting Differences

1 **Fill in the missing words to complete the sentences about classification.**

a) A species is a group of _____ organisms that can interbreed to produce _____ offspring.

b) Individuals of a species show _____. Differences may be _____ (passed on by the parents at fertilisation), for example, blood groups and _____ colour. Differences could also be caused by _____ factors (the conditions they live in).

c) Organisms with similar _____ can be sorted into groups. This grouping is called _____. The first five major groups of living things are called _____.

d) Mammals are covered in _____, give birth to _____ young and _____ the young with mother's milk.

2 **Read the information provided, then answer the questions that follow.**

Pieter and Anjni think that taller people have bigger feet.

They do a survey of their class and produce a scattergram of their results (opposite).

a) Was their idea correct?

b) i) Which point (A–F) on the scattergram shows a person who is tall with large feet?

   ii) Which point (A–F) on the scattergram shows a person who is short with small feet?

c) i) Copy the graph and draw a line of best fit on it.

   ii) What can you say about the people nearest to this line?

   iii) What can you say about the people furthest from the line?

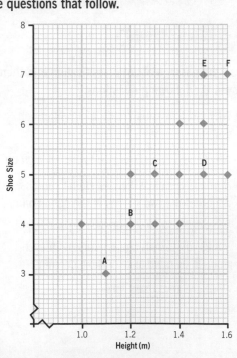

Tom and Beth investigated how pupils in their class were the same and different. They measured the length of the little finger on each pupil's right-hand.

They recorded their results in this table:

Length of Little Finger	Number of Pupils
50 to 55	6
56 to 60	10
61 to 65	6
65 to 70	6
70 to 75	8

1. What unit of measurement did they use?

2. Plot these results as a bar graph.

3. a) What mistake did they make in the way they grouped the finger length results?
   b) What grouping should they have used?

4. Why should each pupil keep their little finger straight while it's being measured?

5. What conclusion can you draw from these results?

6. Do you think that these differences are genetic or caused by their environment?

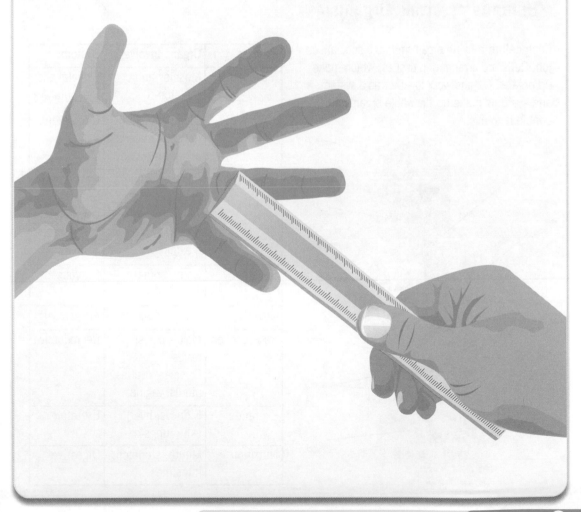

Food and Life

What is Life?

Life is being able to fulfil a series of processes that can be remembered by 'Mrs Gren':

- **M**ovement.
- **R**espiration (release of energy).
- **S**ensitivity.
- **G**rowth.
- **R**eproduction.
- **E**xcretion (removal of waste products).
- **N**utrition.

To be considered alive, an organism must be able to carry out all these processes.

Humans as Living Organisms

Each cell in a living organism has a specialised job. Cells are arranged in tissues, which make up organs. Organs work together as a system, and systems make up the whole organism to enable it to live.

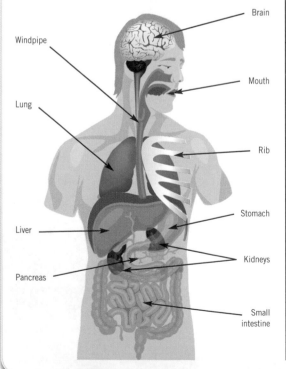

Brain
Windpipe
Mouth
Lung
Rib
Stomach
Liver
Kidneys
Pancreas
Small intestine

Process	Organs Involved	Systems
Movement	Muscles and skeleton	Skeletal and musculature
	Blood vessels transport materials around the body	Circulatory
Respiration	All cells respire, but lungs and windpipe are needed for breathing	Respiratory (or pulmonary)
Sensitivity	Brain; spinal cord; nerves	Nervous
Growth	Cells and organs	All
Reproduction	Male – penis; testes Female – ovaries; uterus; vagina	Reproductive
Excretion	Kidneys; liver; skin; lungs	Excretory
Nutrition	Mouth; stomach; intestines	Digestive

How Food is Used

Your body mainly needs food for...
- energy
- growth
- repair.

You need to have the right amount of each of the seven food types (nutrients) for your individual needs. This is called a **balanced diet**.

Along with enough exercise, a balanced diet will help you to stop becoming overweight and unhealthy.

A baby needs a lot of food for growth, especially for bones and teeth.

An old man needs less food as he's not so active.

An active child needs a lot of food for energy and growth.

Food Types

Nutrient	Needed for...	Found in...
Carbohydrates (starches and sugars)	energy	potatoes; bread; fruit; cereals
Proteins	growth and repair of body tissues	meat; fish; eggs; nuts
Vitamins	helping processes in your body work efficiently, e.g. vitamin D for bone development	citrus fruit; fresh vegetables
Minerals	helping to produce body chemicals and materials, e.g. iron for blood; calcium for teeth and bones	milk; fresh vegetables; fruit
Fats and oils	energy and insulation	milk; butter; cheese; olives
Fibre (roughage)	helping to push food through the body. It's not digested but prevents many bowel problems	wholemeal bread; bran; fresh vegetables; fruit
Water	all processes in your body, which is why you're made up of about 70% water	all foods and, of course, you can drink it!

Food and Life

The Digestive System

The digestive system breaks down the large, insoluble molecules in carbohydrates, proteins and fats into small, soluble molecules that can be absorbed into the blood through the intestine wall. This process is called digestion.

Vitamins, minerals and water are already small enough to pass through the intestine wall, so they don't need to be broken down.

The heart pumps blood around the body so that cells can get the food materials they need.

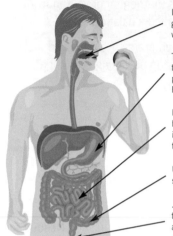

In the mouth, food is ground up and mixed with saliva.

The stomach contains acid to kill bacteria and some protein is broken down here. Food stays here for 2–3 hours.

In the intestines, food is broken down and absorbed into the blood to be taken to the cells that need it.

Undigested food (fibre) is stored for a while...

...until it's removed from the body through the anus as faeces. This is called egestion.

Enzymes

The body produces chemicals called enzymes that break down the large molecules into small ones so they can be absorbed into the blood.

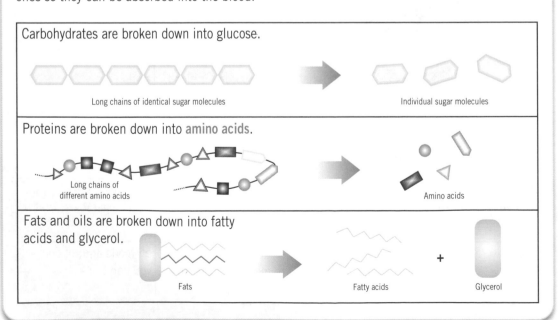

Carbohydrates are broken down into glucose.

Long chains of identical sugar molecules

Individual sugar molecules

Proteins are broken down into amino acids.

Long chains of different amino acids

Amino acids

Fats and oils are broken down into fatty acids and glycerol.

Fats

Fatty acids

+

Glycerol

Photosynthesis

Photosynthesis is the process that green plants use to get their food. It involves...

- plants using the energy from sunlight to turn **carbon dioxide** from the air, and water from the soil, into glucose in the leaf
- oxygen being released into the air as a waste product.

The leaves carry out photosynthesis. They're thin, have a large surface area and are arranged on a branch to get as much light as possible.

The palisade cells are packed with chloroplasts that contain the green pigment, **chlorophyll**, which absorbs the light energy for the plant.

Carbon dioxide is taken from the air through the stomata (pores in the lower epidermis). Water is taken from the soil by root-hair cells and then taken up the stem to the leaf.

The glucose is usually stored as starch in the leaves where it can be used...

- in respiration to release energy for the plant
- to make new cellulose for cell walls
- to make starch for storage in roots or tubers
- to make protein for growth (this needs elements from the soil, such as **nitrogen** and phosphorus, taken in by root-hair cells)
- to make oils for storage in seeds or fruits.

The waste oxygen is released to the air from the stomata.

Photosynthesis is very important. Without it, there would be no food or oxygen.

Leaf Structure

Waxy cuticle
Upper epidermis
Palisade layer
Spongy mesophyll
Lower epidermis
Guard cell
Stomata
Veins

A Root-hair Cell

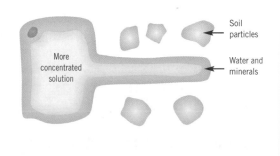

Soil particles
More concentrated solution
Water and minerals

Quick Test

1. Name three processes that all living things must do.
2. The nervous system consists of which three parts?
3. Which food type isn't digested but helps food move through the digestive system?
4. What is the name of the green pigment in plants?

Food and Life

Key Words Exercise

Match each key word with its meaning.

Amino acids	A substance in the diet used for growth and repair of body tissues
Balanced diet	The process by which plants make food
Carbon dioxide	The green pigment that absorbs light energy
Chlorophyll	An element necessary for plants to make protein
Digestion	Helps push food through the intestine and prevents bowel problems
Enzymes	The small molecules produced after protein digestion
Fibre	Chemicals that break down large, insoluble molecules into small, soluble molecules
Nitrogen	Contains the right proportions of the seven food types for you
Photosynthesis	The breakdown of large food molecules into small molecules so they can be absorbed
Protein	A gas that plants need in order to photosynthesise

Comprehension

Read the passage about the model gut, then answer the following questions.

Amylase is an enzyme that breaks down starch into sugar. It's found in saliva. In the experiment shown, some starch solution and amylase were placed in Visking tubing, which was sealed at one end and placed in a beaker of water. Samples of the contents of the tubing were taken every 30 seconds and tested for starch and sugar. Similarly, samples of the water around the tubing were taken every 30 seconds and also tested for starch and sugar. The results showed that sugar appeared in the water outside the tubing after one minute but starch never appeared. Inside the tubing, the solution tested positive for starch immediately and positive for sugar after 30 seconds. However, after three minutes and 30 seconds the contents of the tubing no longer tested positive for starch.

1. Which part of the digestive system is the Visking tubing supposed to represent?

2. Why was there no sugar at first inside the tubing?

3. Which part of the human body is the water in the beaker supposed to represent?

4. What caused sugar to appear in the tubing?

5. Why did sugar appear in the water outside the tubing but starch didn't appear?

6. Why did the starch inside the tubing eventually disappear?

7. How does the experiment compare with what happens in the human gut?

Testing Understanding

1 **Fill in the missing words to complete the sentences about food and life.**

a) A _____ diet provides the appropriate amounts of all _____ major food types.

b) You need _____ and fats for energy. You need _____ for growth and repair of tissues. Minerals and _____ are needed in very small amounts to help your body work well and keep you healthy.

c) If you eat too much food and don't do enough exercise, you may become _____. You need quite a lot of _____ in your diet as 70% of your body is made up of it.

d) Digestion breaks down large _____ into smaller ones using chemicals called _____. This means that food is easily _____ through the intestine wall and into the _____, which carries it to where it's needed.

e) Plants don't eat food from the soil but make their own by _____. The gas, _____, from the air, and _____ from the soil, are turned into glucose using energy from _____. This process also releases _____ into the air as a waste product.

2 **Read the information provided, then answer the questions that follow.**

The table opposite shows the energy requirements of seven different people.

a) Plot this data as a bar chart on graph paper.
b) Suggest why the 21-year-old man needs more energy than the 21-year-old woman?
c) Explain the difference in energy needs between the two 35-year-old men.
d) Explain why the woman who is pregnant needs more energy than the other woman.

Person	Energy Needs per Day (kJ)
6-year-old child	7 000
14-year-old boy	15 500
21-year-old man	13 500
21-year-old woman	11 000
21-year-old pregnant woman	12 500
35-year-old male clerical worker	12 500
35-year-old male manual worker	19 000

Food and Life

Phil and Cherry decided to investigate the effect of light intensity on the rate of photosynthesis in Canadian pondweed.

They set up the experiment as shown and recorded their results in the table below.

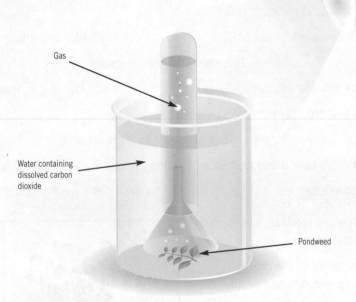

Gas

Water containing dissolved carbon dioxide

Pondweed

Distance from Lamp (cm)	10	20	30	40	50	60	80	100
Amount of Gas Collected in 10 minutes (cm^3)	100	75	65	50	30	15	15	15

1 Plot the data onto a graph grid, labelling the axes clearly.

2 What gas would you expect to collect in the tube?

3 What conclusion can you draw from this graph?

4 How much gas would be collected at:

  a) 25cm?
  b) 120cm?

5 Which is the dependent variable (the variable that's being measured) in this investigation?

6 Why do you think the amount of gas didn't change between 60cm and 100cm?

7 Which key variables would need to be carefully controlled in order to make this a fair test?

Responding to the Environment

Sensing and Doing

Every action you take is a **response** to a **stimulus**, for example...

- you prick your finger on a drawing pin (stimulus) and immediately pull your finger away (response)
- you put food into your mouth (stimulus) and produce saliva (response).

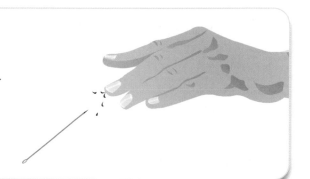

Nerve Impulses

Sensory nerves have **receptors** at one end that transform energy from one type to another. For example, chemical energy is transformed into electrochemical energy in the nerve cells.

An impulse then passes along to relay nerve cells (in the brain or spinal cord) and to **motor nerve** cells, which cause an action, like the movement of a muscle.

In this way, you respond to a variety of stimuli, (for example, smell, taste, movement, light, sound, heat and pressure, etc). The brain coordinates all these responses so that you show patterns of **behaviour**.

By learning, you can change your behaviour. For example, if you learn that a fire is hot then you try not to touch it. But very young babies don't know this, so a fire must be guarded to stop them touching it.

All living things have senses, for example...

- woodlice will move into dark, damp areas
- maggots will move away from light
- young seedlings will grow towards the light (even though plants don't have nerves).

Responding to the Environment

Moving

Very often, your response to a stimulus is to make a movement. To move your bones, you need muscles.

Muscles can...
- **contract** (get smaller)
- relax (return to their original state)
- only exert a force when contracting, which means they can pull but not push.

The muscles in your arm work in opposition to each other, so when one contracts, the other relaxes. This is an example of **antagonistic** action:
1. To move the forearm up, the biceps must contract and the triceps relaxes.
2. To move the forearm back again, the biceps must relax and the triceps contracts.

1

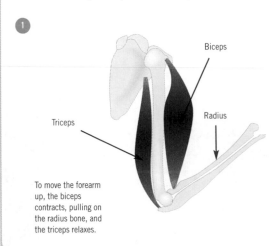

Biceps

Radius

Triceps

To move the forearm up, the biceps contracts, pulling on the radius bone, and the triceps relaxes.

2

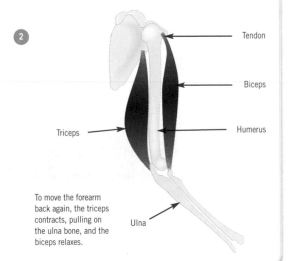

Tendon

Biceps

Humerus

Triceps

To move the forearm back again, the triceps contracts, pulling on the ulna bone, and the biceps relaxes.

Ulna

Joints

Bones don't bend, so in order for you to move easily they need joints. Some joints allow more movement than others, for example...
- the elbow and knee joints are hinge joints and only allow an up-and-down motion
- the shoulder and hip are ball-and-socket joints, which allow up, down and circular motion.

All joints need antagonistic pairs of muscles in order to work efficiently.

Knee Joint   Shoulder Joint

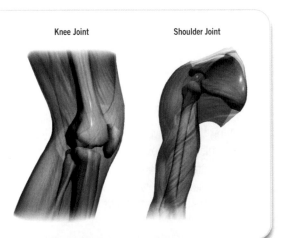

Respiration

In order to move, muscles need energy. Energy is released during **respiration**.

Aerobic respiration is the process by which **oxygen** and **glucose** are combined to release energy for muscle and other cells to use. The equation is:

$$glucose + oxygen \longrightarrow carbon\ dioxide + water + energy$$

This is how aerobic respiration works:

- The glucose molecule contains a lot of energy. You get it from the digestion of carbohydrates.
- The oxygen is taken from the air in the lungs.
- Inside cells, the oxygen is combined with the glucose and this results in the glucose molecule being broken up. **Carbon dioxide** and water are waste products of this energy-releasing process.
- The more work a cell does, the more energy it needs, so when a muscle is working hard it needs more glucose and oxygen.
- This process also generates heat.

A Working Muscle Cell

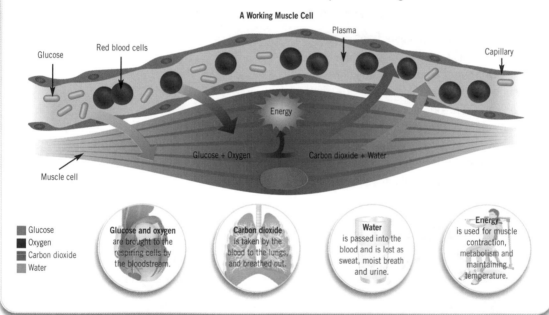

- ■ Glucose
- ■ Oxygen
- ▤ Carbon dioxide
- ▦ Water

Glucose and oxygen are brought to the respiring cells by the bloodstream.

Carbon dioxide is taken by the blood to the lungs, and breathed out.

Water is passed into the blood and is lost as sweat, moist breath and urine.

Energy is used for muscle contraction, metabolism and maintaining temperature.

Respiration Around the Body

Respiration is a process that goes on in every cell in the body to release energy from food.

The blood is very important because it carries oxygen from the lungs and glucose from the intestines to the cells for respiration.

Blood also carries away the carbon dioxide dissolved in the plasma (blood liquid) to the lungs. Water is taken to the kidneys so that it can be **excreted** (got rid of) from the body.

When you exercise, your heart and lungs have to work harder to get oxygen and glucose to the cells.

If the muscle cells don't get enough glucose or oxygen, they may become fatigued (tired) and stop working properly, which may result in cramp in the muscle. Cramp is where the muscle contracts and doesn't relax again, causing pain.

Responding to the Environment

Innate and Learned Behaviour

Behaviour in humans and animals is **either...**
- innate (instant and caused by genes) **or**
- learned (the result of experience).

The simplest type of innate behaviour is a reflex action like immediately pulling your hand away from a hot or sharp object. This behaviour protects you from harm. Some innate behaviour can be altered by experience.

These are examples of innate behaviour:
- Some animals, like woodlice, move very fast when in the open and only slow down when they're under something. This helps to keep them safe from animals or birds that might eat them.
- Mosquitoes bite and obtain food (blood) from exposed flesh if that person's body temperature attracts them.

Learned behaviour is developed through experience. Some animals start learning as soon as they're born and they will follow the first object they see that has certain characteristics.

For example, chicks will follow their mother and some geese will follow wellington boots worn by their owner if they see them first as soon as they hatch.

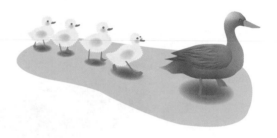

You can become used to a stimulus that offers no danger. For example, if you enter a room with a bad smell, at first you react. After a while, the smell 'goes' because the body realises that it's not a danger and so 'switches off' the response.

Quick Test

1. What is an action you take as a result of a stimulus called?
2. What is the term for a pattern of responses to a variety of stimuli?
3. When do muscles exert a force?
4. What do you call two muscles that work in opposition to each other?
5. Which gas does aerobic respiration use to help break down glucose?

KEY WORDS
Make sure you understand these words before moving on!
- Antagonistic pair
- Behaviour
- Carbon dioxide
- Excretion
- Glucose
- Innate behaviour
- Learned behaviour
- Motor nerve
- Muscle contraction
- Oxygen
- Receptor
- Respiration
- Response
- Stimulus

Key Words Exercise

Match each key word with its meaning.

Antagonistic pair	Something that causes a response in the nerve cell
Behaviour	An action as a result of a nervous impulse, for example, moving a muscle
Carbon dioxide	A pattern of actions carried out by an animal
Excretion	The end of a sensory nerve that transforms one type of energy into a nerve impulse
Glucose	A type of nerve that causes an action to take place
Innate behaviour	When a muscle shortens to exert a force
Learned behaviour	Two muscles that work opposite to each other
Motor nerve	A combination of chemical reactions that release energy in a cell
Muscle contraction	Removal of waste products of chemical reactions from the body
Oxygen	The energy-rich molecule used in respiration
Receptor	A gas that's a waste product of respiration
Respiration	A gas used up to help break down glucose in respiration
Response	Behaviour that's instant and caused by genes
Stimulus	Behaviour that occurs as a result of experience

Comprehension

Read the passage about Ivan Pavlov, then answer the following questions.

1. Why did the dogs salivate when given food?

2. What were the two stimuli that were associated by the dogs?

3. Why did Pavlov have to use other dogs for this experiment?

4. How would this behaviour be of advantage to the dogs?

5. Describe a conditioned behaviour you might show.

In the late 1890s, a scientist called Ivan Pavlov noticed that the dogs he was studying would salivate when they knew they were about to be fed. He noticed that the dogs were able to recognise that food was on its way by a bell that rung every time the laboratory door was opened. So, he carried out a set of experiments on other dogs.

First he rang a bell, which caused no response in the dogs. Then, after the bell was rung, he gave the dogs food, which made them salivate. This was repeated several times.

Eventually, the dogs began to respond by salivating to the sound of the bell, even when no food was given. This is learning by association and is known as conditioned behaviour.

Responding to the Environment

1. **Fill in the missing words to complete the sentences about responding to the environment.**

   a) Each action that a body makes is a _____ to a _____.

   The pattern of actions shown is known as _____.

   b) Sensory _____ have a _____ at one end that turns

   one type of energy into a nerve / neurone _____.

   c) When a muscle _____ it exerts a force but when it _____

   it doesn't, so it can only _____ and not push. In order for a bone to

   move one way, it needs one muscle, and to move the other way it needs a second. These

   muscles are found in _____ pairs.

   d) Energy from _____ is needed for a muscle to move. _____

   helps to break down the energy-rich molecule _____ to release energy to

   the muscle cell. If the muscle doesn't get enough energy then _____ sets

   in and it can't work efficiently.

   e) The behaviour shown as a result of all these actions is either caused by genes and is

   _____ or is a result of experience, when it's _____.

   It all adds up to make you unique in the way you behave.

2. **Study the diagram opposite showing a reflex pathway in a human, then answer the following questions.**

   a) Label the receptor on the diagram.
   b) Label the motor neurone and sensory neurone on the diagram.
   c) Suggest a stimulus to the hand that might start a reflex response and describe the possible action.
   d) Draw arrows on the diagram to show the direction of the path taken by the nervous impulses.

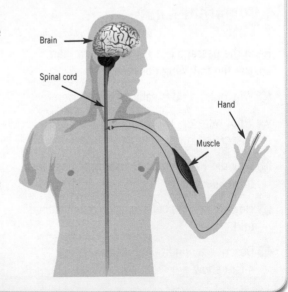

Brain

Spinal cord

Hand

Muscle

Terry and Trudi decided to find out the effect of temperature on the respiration rate of woodlice.

They set up the experiment as shown and carried it out in a range of temperatures. They took great care not to harm the animals as they did this.

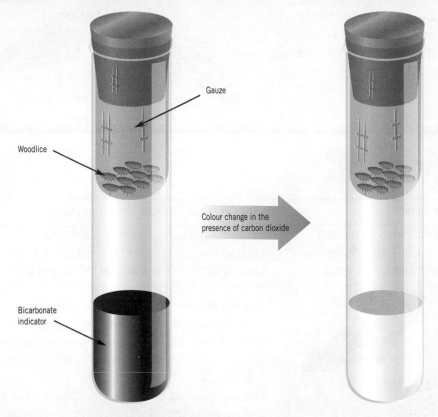

Gauze

Woodlice

Colour change in the presence of carbon dioxide

Bicarbonate indicator

They recorded their results in the table below.

Temperature (°C)	5	10	15	20	25	30	35	40
Time Taken to Turn Indicator Yellow (mins)	35	30	28	20	12	10	8	5

1. Plot the data onto a graph grid (label the axes clearly).

2. What conclusion can you draw from this graph?

3. Why do you think this pattern was shown?

4. Which is the independent variable (the variable that's being changed) in this investigation?

5. Why didn't Terry and Trudi increase the temperature higher than 40°C?

6. Which key variables would need to be carefully controlled in order to make this a fair test?

Survival in the Environment

Fitting into the Environment

In order to survive, each organism must fit into its environment as best as it can. Within a species, each organism can look very similar but there can still be differences between each other. This is called variation.

Examples of Variation

Organism and Variation	How the Organism Fits into the Environment
The peppered moth has two distinct variants (versions): one has light-coloured wings and the other has dark wings.	• Before the Industrial Revolution, the light-coloured variant was more common as predatory birds struggled to see it on tree trunks, but the dark variant stood out. • With the arrival of industry and its smoke, the surrounding buildings and some trees became blackened with soot. Then the dark-winged variant was at an advantage because it was hard to spot, but the light ones were easily seen. • Both variants have advantages in different environments. They're still the same species and can interbreed easily, producing offspring of both variant types.
Some moles have larger feet than others.	In an environment where the soil is very hard, the large-footed moles are better able to dig through the soil to find food, like earthworms. In soft soils, neither variant is at any great advantage.
Different foxes have ears of a different size.	In a hot desert, foxes with larger ears find it easier to stay cool because their ears act as radiators, releasing excess heat. As a result, desert foxes with larger ears have an advantage over those with smaller ears.

Genes and Chromosomes

Variation is caused by organisms within a species having different genetic information in their cells:

- The **nucleus** of each cell contains thousands of **genes** arranged on threads called **chromosomes**, like links on a chain.
- The genes are the instructions that control all the **characteristics** organisms have.
- Even though organisms of the same species have a lot of genes in common, making them similar, the small differences create the variation between them.

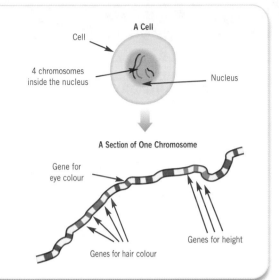

A Cell

Cell

4 chromosomes inside the nucleus

Nucleus

A Section of One Chromosome

Gene for eye colour

Genes for height

Genes for hair colour

Limitations of Variation

In most cases, the differences between organisms within a species are only small, so they don't really have an environmental advantage over each other.

For example, it's unlikely that the differences in the two cats opposite would make any difference to their ability to survive in any environment. Only large differences would create an advantage or disadvantage.

The differences in the genetic make-up of organisms are...

- entirely random and aren't a response to any change in the environment, so it's luck that produces an advantageous difference
- usually small, so it often takes a very long time (i.e. over many generations) for larger differences to develop and by then it may be too late for these changes to be an advantage
- sometimes a disadvantage rather than an advantage, as with the non-camouflaged bird opposite.

Usually, variation makes little difference to the survival of an organism; it simply makes the organism 'individual'.

Survival in the Environment

Genetic and Environmental Factors

Genetic variation can be passed on from one generation to the next, but differences caused by the environment can't be passed on:

- Sex, blood group and the shape of earlobes are **genetic factors** and aren't influenced by **environmental factors**.
- Strength, haircut and the ability to speak a language are the result of environmental and not genetic factors.
- Characteristics like height, personality, skin colour and hair colour can be the result of both environmental and genetic factors.

Identical twins have exactly the same genes because they were formed from the same fertilised egg, but their lifestyle and their experiences can affect them greatly.

Twins have the same genes but environmental factors can change their appearance.

Nature Versus Nurture

How much of the ability to play a sport well is down to genes and how much is down to practice and hard work? Can anyone be a great scientist or write a best-selling novel, or do you have to be born with the right genes?

You will have to try to decide because no-one really knows. This is the nature (genetic causes) versus nurture (environmental causes) debate.

Examples of Survival

Characteristics inherited by one generation from another help organisms to survive in changing environments.

Although these genetic changes may be small, often the changes in the environment are also small. As a result, organisms showing variation survive:

Organism	How Genetic Variation Helps the Organism to Survive
Bluebell	Some bluebells grow earlier than others and are able to capture the light with little competition, allowing them to grow stronger and flower earlier.
Giraffe	A giraffe with a longer neck than others may be able to reach further up the trees to gain food, allowing it to survive better.
Olive tree	Olive trees traditionally like warm climates. But those that can withstand slightly cooler temperatures may be able to survive in Britain as climate change makes the conditions more favourable for them.
Staphylococcus bacteria	Those Staphylococcus bacteria that have resistance to antibiotics (drugs that have come from microbes) have caused problems in hospitals, for example MRSA (methicillin-resistant Staphylococcus aureus) infections. This is because they've been able to survive antibiotic treatment.
Rat	Some rats can survive even when the poison Warfarin has been laid to try to kill them. As a result, other types of poison have to be used.

Quick Test

1. What name is given to the differences in characteristics among organisms of the same species?
2. What is the cause of the differences between these organisms?
3. What is the name of the thread-like structures found in a nucleus?
4. Apart from inherited characteristics, what other factors may affect how a person looks?

KEY WORDS

Make sure you understand these words before moving on!

- Antibiotics
- Characteristics
- Chromosome
- Environmental factors
- Genes
- Genetic factors
- Inheritance
- MRSA
- Nucleus
- Resistance
- variant
- variation

Survival in the Environment

Key Words Exercise

Match each key word with its meaning.

Antibiotics	•	• Differences within a single species
Characteristics	•	• An organism that shows a certain difference in a species
Chromosome	•	• The part of the cell where the genetic material is found
Environmental factors	•	• Units that determine an organism's characteristics
Genes	•	• Those features that can be seen in or on an organism
Genetic factors	•	• A disease often found in hospitals, caused by a bacterium
Inheritance	•	• Features of the surroundings that may affect an organism
MRSA	•	• Influences on the genetic material of an organism
Nucleus	•	• The passing on of characteristics from one generation to the next
Resistance	•	• Drugs, produced from microbes, that are used to control bacterial infections
Variant	•	• The ability to be relatively unaffected by something, for example, a drug
Variation	•	• A thread-like structure, like a string of beads

Comprehension

Read this description of David, then answer the following questions.

1. Which of David's features do you think are inherited?

2. Which features do you think are affected by the environment?

3. Which features may be a result of both the environment and genes?

David has brown hair, in which he has dyed blond highlights, and green eyes. He is quite tall and quite skinny, but he can run fast and is the star member of the local basketball team. He recently went on a basketball tour in southern Italy and his skin has turned quite pink as a result.

His eyebrow is pierced and he has a scar on his right knee where he fell off his bike when he was very young. He has a straight nose and large ears that stick out.

Testing Understanding

1 Fill in the missing words to complete the sentences about survival in the environment.

  a) Variation is differences in _____ between individuals.

  b) Characteristics may be passed on from parents to their offspring in their
_____ . These are _____ factors.

  c) Characteristics that are affected by our surroundings are _____ factors.

  d) Differences between individuals of the same _____ may allow them to
_____ in an environment that's changing.

  e) Some features are a result of both types of factor and many people are unsure of just how
much of each type is really responsible. This argument is called the _____
versus _____ debate.

2 Study the boxes below about why dinosaurs became extinct millions of years ago, then answer
the questions that follow.

Here is some information about dinosaurs and plants: • Most dinosaurs ate plants. • Most dinosaurs were the largest animals alive at that time. • Plants can't grow without the Sun. • Dinosaurs couldn't survive very cold conditions.	Here are some ideas about why the dinosaurs became extinct: • A new disease killed them. • A new predator killed them. • New animals ate all their food. • A huge meteorite hit the Earth, resulting in dust blocking out the Sun.

  a) Using this information, choose an idea that you think isn't very likely to have happened.
Give a reason for your answer.

  b) Explain how a meteorite could have been the cause of the dinosaurs' extinction.

Survival in the Environment

Fiona and Nigel carried out an experiment to see if the camouflage of an insect helped to prevent it from being eaten by birds.

To do this, they made 'insects' from bread and used food colourings to dye them different colours.

Then they placed equal numbers of coloured 'insects' on different pieces of coloured card and placed them randomly in the school garden.

They hid and watched the garden for two hours, then counted the 'insects' left.

Their results are shown in the table below.

1. What conclusions can you make from these results? Give reasons for your answer.

2. Why did Fiona and Nigel have to watch the garden during the investigation?

3. How could they improve this investigation?

4. What factors did they need to try to control in this investigation?

Colour of 'Insect'	Colour of Card	Number of 'Insects' at Start	Number of 'Insects' After Two Hours
Red	Red	10	8
Yellow	Yellow	10	7
Blue	Blue	10	6
Red	Yellow	10	1
Yellow	Blue	10	6
Blue	Red	10	2
Red	Blue	10	0
Yellow	Red	10	6
Blue	Yellow	10	2

Environmental Relationships

Describing an Environment

These terms help you to describe an environment:
- Habitat – the particular type of area in which an organism lives.
- Population – the total number of individuals of the same species that live in a habitat.

- Community – all the different organisms living in a habitat (the total of all populations).
- Ecosystem – the total community of living things, together with all the physical features (rainfall, temperature, wind, water, light, etc) in the habitat.

Types of Interaction Within Environments

All the organisms that live in a particular habitat will be adapted to it. They will all show different characteristics that help them to survive and breed to maintain the population size.

As a result, the biodiversity (variety of life) found in different habitats and ecosystems is very different.

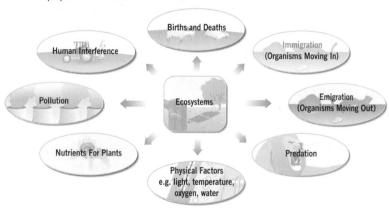

Size of Populations

The distribution (spread) and abundance (numbers) of organisms in a habitat depend on…
- the adaptation of the organism to the surrounding environment
- how much competition there is (from other species and members of the same species)
- the number of births and deaths (from predation, disease, etc).

How many new organisms are produced and how many are removed

Environmental Relationships

Collecting Habitat Data

The best way to collect the correct data is to count everything and measure everything. But this is never really practical, so you have to sample a habitat and use the sample to estimate the totals.

Estimating Populations

Once the animals have been trapped and counted, they can be marked in some way, for example, with a spot of paint. The organisms can then be released.

After a time, another sample can be trapped and a mathematical formula used to estimate the population size. This is called a 'capture-recapture' technique.

For plants and some very slow animals, you can use **quadrats**. Quadrats are usually squares (but they can be any shape) of a known area:
- They're randomly placed in the habitat.
- The numbers of organisms found in each quadrat are counted and the total data is used to estimate the population.

Remember not to harm any organisms and return everything afterwards.

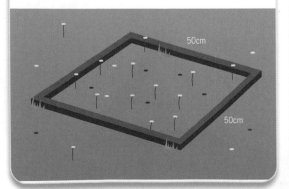

Identifying Organisms

You can look at plants and trap animals in pitfall traps, nets and pooters (a device that enables you to suck insects into a collection chamber without harming them). You can use keys to look up what the organisms are, then count and record the numbers.

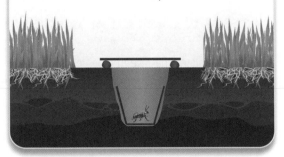

A Pitfall Trap

Measuring the Conditions

You can use different sorts of meters and data loggers to measure soil, water and air conditions, for example...
- a thermometer to measure temperature
- a light meter for light levels
- a flow meter for water flow
- a pH meter or universal indicator to measure the pH of soil.

This is preferably done over a period of time, not just in one go.

Food Chains

Once all the data has been collected, you can use the information to make **food chains** and food webs to show the feeding relationships in the habitat, for example:

Wheat   Harvest mouse   Owl

Energy in Food Chains

This is how energy would be passed along the food chain above:

- The wheat plants get their energy from sunlight. They use this energy for photosynthesis to make their own glucose. They also take in nutrients from the soil and use this to build their own proteins and fats, etc.
- Plants use up some of their glucose in respiration, which causes thermal (heat) energy to be given out into the surroundings. So, not all the energy produced by the plant is passed onto the harvest mouse.
- After eating the wheat, the mouse also respires, releasing heat. Some of this energy is used as kinetic (movement) energy. Some of the energy passes out from the mouse in urine and faeces. So, not all the energy from the mouse is passed onto the owl, and so on.

Numbers of Organisms

A lot of energy present in one step of the food chain isn't available to the animal in the next step. As a result...

- the number of organisms decreases dramatically from one feeding level to the next
- food chains are usually only three, four or five steps. There isn't enough energy left at the top to support another step.

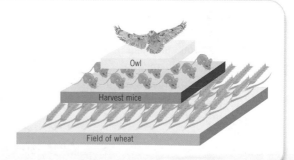

Owl

Harvest mice

Field of wheat

Environmental Relationships

Environmental Pyramids

Pyramid diagrams…
- help to show the numbers of producers (green plants), primary consumers (herbivores) and secondary consumers (first-level carnivores)
- give you an idea of how much energy is available at each feeding level and how it decreases as the food chain gets longer
- help you to compare different food chains to look at similarities and differences.

You can use a food chain and the number of organisms to produce a pyramid of numbers:
- The width of each bar shows the number of organisms and you usually get a typical pyramid shape ①.
- Occasionally you get a strange-looking pyramid. For example, pyramid ② is different because there is only one oak tree but many caterpillars eating its leaves. What this type of pyramid doesn't allow for is size.

In a pyramid of biomass, all the organisms are weighed and the total mass of each type of organism is calculated. The mass of the oak tree is far greater than the mass of all the caterpillars, so you get a typical pyramid shape again ③.

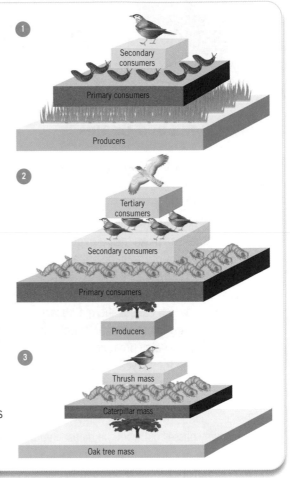

Quick Test

1. What is a habitat?
2. What is a population?
3. What do you call small squares used to sample an area?
4. Why is the energy available to the next step in a food chain less than at the beginning?
5. What is biomass?

KEY WORDS

Make sure you understand these words before moving on!
- Biodiversity
- Community
- Distribution
- Ecosystem
- Food chain
- Habitat
- Immigration
- Population
- Primary consumers
- Producers
- Pyramid of biomass
- Pyramid of numbers
- Quadrat
- Secondary consumers

Key Words Exercise

Match each key word with its meaning.

Biodiversity	The particular type of area in which an organism lives
Community	The total number of individuals of the same species in a habitat
Distribution	The total number of populations in a habitat
Ecosystem	The community of organisms together with the physical conditions in a defined area
Food chain	To move into an area
Habitat	The range of different species in an area
Immigration	The spread of individuals of a species in an area
Population	A device for sampling 'fixed' organisms in a large area
Primary consumers	A sequence of organisms showing 'what eats what' in an area
Producers	A way of showing the relative numbers of organisms in a food chain or web
Pyramid of biomass	A way of showing the relative mass of organisms in a food chain or web
Pyramid of numbers	A name given to green plants that can make their own food
Quadrat	A name given to animals that eat producers
Secondary consumers	A name given to animals that eat herbivores

Comprehension

Read the passage about food chains, then answer the following questions.

1 What is the name of the process by which producers convert light to chemical energy?

2 Why are rainforests more productive than deserts?

3 What are animals that eat producers called?

4 What are animals that eat other animals called?

5 How is most energy lost from each step of the food chain?

Sunlight energy sustains almost all the approximate two million species of organism on Earth. Producers can turn this light energy into chemical energy in the form of glucose.

Different ecosystems produce different amounts of chemical energy because they vary in the amount of light, water and nutrients available to them, and in factors such as temperature. Tropical rainforests are highly productive, while deserts are unproductive.

Animals eat the producers and, in turn, are eaten by other animals. In the process, energy that's stored in the bodies of each organism flows along a food chain. At each step, some energy is lost as heat in respiration, as kinetic (movement) energy and through excretion in urine and faeces. As a consequence, food chains rarely go beyond four or five steps.

Environmental Relationships

Testing Understanding

1 Fill in the missing words to complete the sentences about environmental relationships.

a) Animals and plants _____ to their environment, which increases their chances of _____.

b) All the organisms living in a particular area are called a _____.

c) You can use fieldwork to study the _____ of organisms in a _____.

d) You can _____ a large area using a _____ to estimate the numbers of organisms.

e) A pyramid of _____ shows the total number of each organism in a food chain. A pyramid of _____ shows the mass of each organism in a food chain. This gives you some idea of the amount of _____ at each stage.

f) The total of all the organisms and the physical factors affecting an area is referred to as an _____. You can measure the physical factors using instruments like a _____ for temperature or a _____ _____ for light levels.

g) The number of animals can be estimated by using a method called capture- _____.

2 Study the environmental pyramids for the food chain below.

grass ➡ rabbit ➡ fox

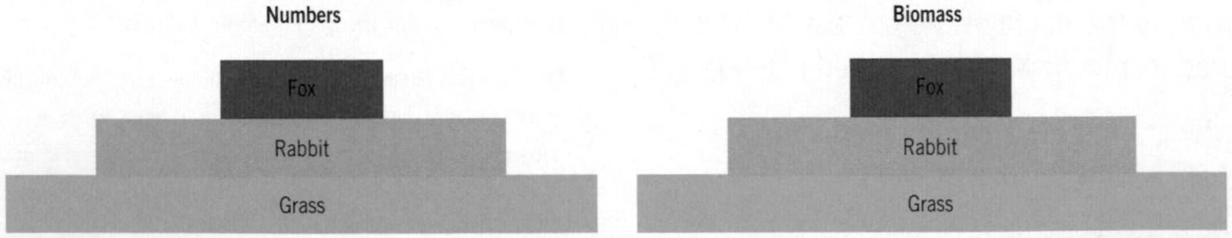

Numbers

| Fox |
| Rabbit |
| Grass |

Biomass

| Fox |
| Rabbit |
| Grass |

Now draw a pyramid of numbers and a pyramid of biomass for each of the following food chains:

a) broad bean ➡ blackfly ➡ blue tit ➡ sparrowhawk

b) dandelion ➡ mouse ➡ stoat ➡ fleas on stoat

A group of pupils used a net to catch stonefly nymphs at different points along a river. They began the sample on the edge of town.

Stonefly nymphs prefer water with high oxygen levels (unpolluted), so the pupils predicted that there would be more nymphs further away from the town.

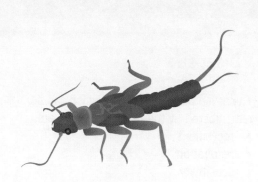

The pupils used exactly the same method of capture at each point and took great care not to harm the animals as they did this. They recorded their results in the table below.

Sample Distance (metres from edge of town)	0	50	100	150	200	250
Number of Stonefly Nymphs	12	22	28	34	48	42

1 Plot the data onto a graph grid, labelling the axes clearly.

2 Which variables would the pupils have to assume would remain the same in order to make this investigation a fair test?

3 Which is the independent variable in this investigation?

4 Do these results support the pupils' prediction? Give a reason for your answer.

5 The result at 250 metres was lower than at 200 metres. Can you suggest a reason for this?

6 Why might the town make a difference to the number of stonefly nymphs?

Disruption of Life Processes

Disease

The seven processes of life, which can be remembered by the term 'Mrs Gren', are...

- **m**ovement
- **r**espiration
- **s**ensitivity
- **g**rowth
- **r**eproduction
- **e**xcretion
- **n**utrition.

A variety of factors can stop these seven processes working efficiently. If any of the processes are disrupted, then you can suffer from illness or disease.

Biological Factors

Microorganisms (microbes) like bacteria, fungi and viruses can affect the way your body works. Disease-causing microbes are called **pathogens**.

Pathogen Type	Examples and Effects
Bacterial	<ul><li>Some forms of meningitis – meningitis affects the membranes in the brain and this can cause brain cells to die, therefore disrupting many of the life processes.</li><li>Weil's disease – caught from contact with rat urine and can affect the excretory system by damaging the liver and kidneys.</li><li>Syphilis – causes ulcers on the reproductive organs that can spread to other areas of the body.</li></ul>
Viral	<ul><li>Influenza ('flu') – causes sore throats and headaches, affecting respiration.</li><li>HIV (Human Immunodeficiency Virus) – attacks the immune system and can allow any infection to damage any part of the bodily processes.</li><li>Mumps – causes the swelling of salivary glands, which affects the digestive system.</li></ul>
Fungal	<ul><li>Ringworm – affects the growth of the skin, appearing like a raised reddish ring on the skin.</li><li>Athlete's foot – affects movement by causing itchy, flaking skin, especially on feet and toes (although it can affect other areas).</li></ul>

Smoking and the Lungs

Tobacco smoke contains chemicals that affect several bodily processes. The lungs are particularly affected.

Oxygen from the air and carbon dioxide from the blood are exchanged in the lungs. The oxygen is used for respiration and the carbon dioxide is the waste gas produced.

The lungs are adapted for gaseous exchange by having a very large surface area and thin walls that are near to many blood capillaries. This allows the gases to diffuse in and out of the blood.

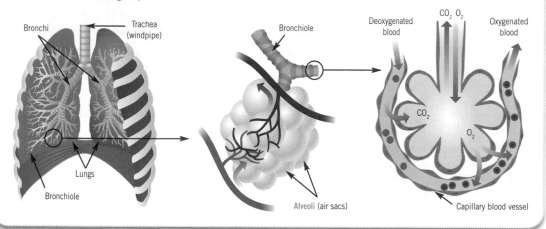

Bronchi

Trachea (windpipe)

Bronchiole

Lungs

Bronchiole

Alveoli (air sacs)

Deoxygenated blood

CO_2 O_2

Oxygenated blood

CO_2

O_2

Capillary blood vessel

The Effects of Smoking

Tobacco smoke disrupts the way in which organs, especially the lungs, work.

You can see the brown tarry deposit left if the smoke is passed through cotton wool in the apparatus shown opposite.

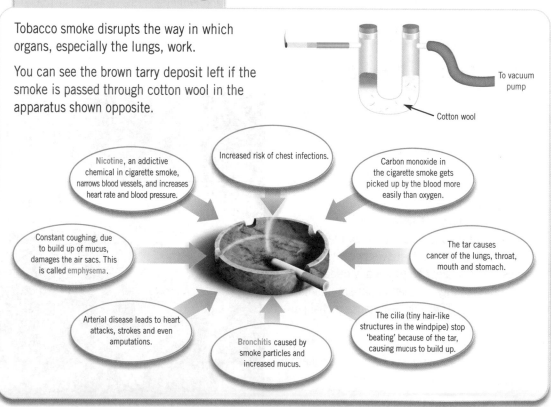

To vacuum pump

Cotton wool

Nicotine, an addictive chemical in cigarette smoke, narrows blood vessels, and increases heart rate and blood pressure.

Increased risk of chest infections.

Carbon monoxide in the cigarette smoke gets picked up by the blood more easily than oxygen.

Constant coughing, due to build up of mucus, damages the air sacs. This is called emphysema.

The tar causes cancer of the lungs, throat, mouth and stomach.

Arterial disease leads to heart attacks, strokes and even amputations.

Bronchitis caused by smoke particles and increased mucus.

The cilia (tiny hair-like structures in the windpipe) stop 'beating' because of the tar, causing mucus to build up.

Disruption of Life Processes

The Effects of Alcohol

In moderation, alcohol is relatively harmless. But alcohol abuse and dependency can cause serious problems:

- Liver damage – alcohol is a mild poison and causes parts of the liver to become fibrous (turn into fibres) and therefore useless.
- Brain damage – regular doses of alcohol lead to increased brain-cell death and a drop in mental performance, e.g. memory.
- Impaired judgement – this occurs while under the influence of alcohol. Acts of bravado or foolhardiness can have fatal effects.
- Addiction – extreme alcohol dependency can lead to regular days off, reduced performance at work, violence and money problems.

The Effects of Solvents

Inhaling the vapour from certain household substances, such as glues and paints, can cause problems:

- Hallucinations – users may lose their grip on reality.
- Personality change – users may start to display different personality traits.
- Damage to organs (e.g. the lungs, brain, liver and kidneys) – this damage is usually permanent.

The Effects of Other Drugs

Other drugs can also affect bodily processes:

- **Hallucinogens** (e.g. Ecstasy, L.S.D.) – these cause hallucinations and, in the case of Ecstasy, the feeling of being full of energy, which can lead to dehydration and collapse.
- **Depressants** (e.g. alcohol, barbiturates) – these depress the nervous system.
- **Stimulants** (e.g. amphetamine, methedrine, cocaine and crack) – users become psychologically and physically dependent on the feelings of energy that stimulants can cause. They can also cause a change in personality.
- **Pain killers** (e.g. heroin, morphine) – these can cause addiction and complete collapse of personality and self-discipline.

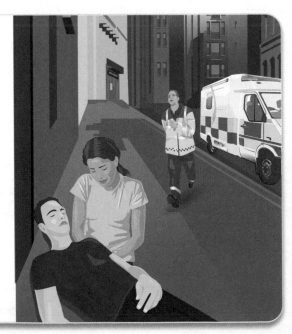

Natural Barriers to Disruption

Your body has natural defences that help to stop processes being disrupted:

- Your skin is slightly acidic (pH 5.5), which will kill many microbes. The skin is also a physical barrier.
- You produce sweat, tears and ear wax, which can all destroy microbes.
- If you get a cut, blood clots seal the wound.
- Your windpipe (trachea) is lined with mucus, which traps microbes, and has tiny hairs (cilia) that move the microbes up and into the oesophagus.
- Your stomach contains acid that kills microbes.

If any microbes get past these defences, your white blood cells can...

- dissolve, or engulf and digest, the microbes
- produce antitoxins that work against poisons made by invading microbes
- produce antibodies that stick to the microbes to stop them working.

Once invaded, the body 'remembers' the disease and can make the antibodies again very quickly if needed. This is called immunity.

Antibodies can be passed on to unborn babies through the placenta and to young babies through the breast milk, making them immune too.

How Antibodies Work

Microbe

Antibodies

White blood cell   A white blood cell releasing antibodies, which attack the microbe before it's engulfed by another white blood cell.

Artificial Barriers to Disruption

You can be immunised against a disease:

- You can have an injection of weakened or dead microbes that cause your body to make antibodies. This will protect you in future. An example is the anti-HPV vaccine (against cervical cancer), which is only made up of virus-like particles.

- You can also be injected with the antibodies themselves. These are often produced in laboratory-bred animals. This type of vaccine usually requires regular booster doses and doesn't last for life. An example is the anti-tetanus vaccine.

Quick Test

1. What is a pathogen?
2. What three types of microbe can cause disease?
3. What is the addictive drug in tobacco?
4. What are the air sacs at the end of the bronchioles called?
5. Alcohol affects which two organs the most?
6. What are passed onto babies in breast milk to give them immunity from a microbe?

KEY WORDS
Make sure you understand these words before moving on!

- Alveoli
- Antibodies
- Antitoxins
- Bronchitis
- Emphysema
- Hallucinogens
- Immunity
- Nicotine
- Pathogen
- Vaccine

Disruption of Life Processes

Key Words Exercise

Match each key word to its meaning.

Alveoli	Protection from disease by the production of antibodies
Antibodies	A collection of dead or weakened microbes, or antibodies, injected to give protection
Antitoxins	The air sacs where oxygen enters and carbon dioxide leaves the blood
Bronchitis	Substances produced by white blood cells that can neutralise a specific microbe
Emphysema	An addictive chemical found in cigarette smoke
Hallucinogens	Chemicals that neutralise a poison produced by a microbe
Immunity	A smoking-related disease caused by excessive mucus and smoke particles
Nicotine	A condition in which air sacs are damaged due to constant coughing
Pathogen	A disease-causing microbe

Comprehension

Read the passage about a study into the causes of lung cancer, then answer the following questions.

1. How might advertising be used to create the impression that smoking is 'cool'?

2. Describe several ways in which the 'smoking group' and the 'control group' should have been identical in order to make this a fair test.

3. Why do you think that the results of the study provoked a hostile reaction from the tobacco companies?

4. Comment on the following statement: "Being a smoker means that you will definitely get lung cancer at some stage."

Scientists investigating the increase in deaths from lung cancer during the 1940s and 1950s began to suspect that the underlying cause was related to smoking. At the time, many people assumed that smoking was safe and thought that it was rather sophisticated – an impression that was reinforced through heavy advertising by the tobacco companies.

A major study compared a group of smokers and a group of non-smokers over a long period of time. The study showed that the smokers were more likely to get lung cancer than the non-smokers, and that the more a person smoked, the greater their chances were of getting lung cancer.

The non-smokers in this study were called the 'control group' and were chosen so as to differ from the smoking group in the fact that they had never smoked. The results of the study provoked a hostile reaction from the tobacco companies and came as a shock to many people. However, it took many years to build up public awareness of the dangers of smoking, partly due to the cigarette manufacturers' huge advertising budgets.

Testing Understanding

1 Fill in the missing words to complete the sentences about the disruption of life processes.

   a) There are three types of microbe that can cause disease: _____,

       _____ and _____ .

   b) Meningitis is caused by _____ . Mumps is caused by _____

       and athlete's foot is caused by _____ .

   c) Once microbes are inside the body, you depend on your _____

       _____ cells to 'fight' them. Some produce_____ that

       neutralise poisons; others produce _____ that _____ to

       microbes to prevent them working. Others may dissolve them, or engulf and

       _____ them.

   d) Immunity can be given either by catching the disease or by _____ , where

       weakened or _____ microbes are injected, or by the injection

       of _____ made outside the body.

   e) The body can also be damaged by the abuse of drugs like _____ ,

       which is legal but can damage the liver and brain. There has also been an increase in the

       use of illegal _____ , such as Ecstasy and cocaine. Users can become

       dependent on, or get _____ to, these drugs.

2 Study the table about smoking and lung cancer, then answer the questions that follow.

Year	Lung Cancer Deaths per 100 000 Men	Cigarettes Smoked per Man per Year
1900	10	800
1920	25	2050
1940	50	4000
1960	175	3500
1980	160	3000

   a) Plot the above data onto two separate
       graph grids as line graphs, with the year
       on the *x*-axis.

   b) What conclusion can be drawn from the
       results about the number of cigarettes
       smoked per man per year?

   c) What conclusion can be drawn from
       the results about the number of lung
       cancer deaths?

Disruption of Life Processes

Karen and Daniel set up an investigation into the effects of an antibiotic drug, penicillin, on the growth of bacteria.

First they inoculated a Petri dish containing agar jelly with bacteria and allowed it to grow over the plate. They then placed four discs of blotting paper onto the surface.

Each disc was soaked in a known concentration of the antibiotic. They thought that the antibiotic would kill the bacteria on the plate.

After one week, they looked at the dish and could see areas where the bacteria had disappeared. The pattern they could see was as shown below.

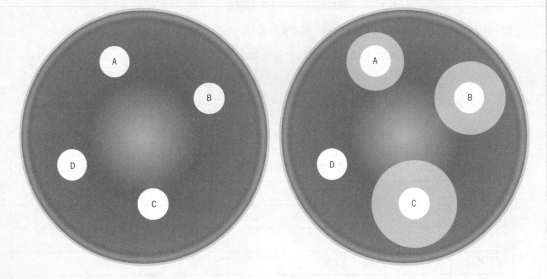

Dish at Start   Dish After One Week

1. Karen and Daniel could see that the area around each disc was different, but how could they measure this?

2. Which disc had the greatest concentration of penicillin on it?

3. Disc D showed no difference. This was because they had only soaked it in water. Why would they do this?

4. They had to carry out this investigation very carefully. Why would they need to be especially careful when doing this?

5. After carrying out each stage of the investigation, both pupils had to wash their hands in antiseptic soap. Why was this necessary?

Variety in the Environment

Sexual Reproduction and Variation

In order to reproduce sexually, organisms must produce specialised cells called gametes:

- Gametes are the 'sex cells' – eggs and sperm in animals, and ovules and pollen in plants.
- The nuclei of these cells contain DNA, which carries the genetic information determining the characteristics that will be inherited.
- When the nuclei fuse at fertilisation, the cell that's produced carries genes from both parents, resulting in a new individual.

Sexual reproduction means that...

- the genetic information from both gametes is passed on to the new individual
- the characteristics of the offspring are similar to the parents, but show quite a lot of differences to them.

Some of the differences between an individual and its parents may be an advantage (e.g. being taller may help in basketball) or a disadvantage (e.g. being taller may make it uncomfortable to drive a small car). Other differences may be neither an advantage nor a disadvantage.

Fertilisation

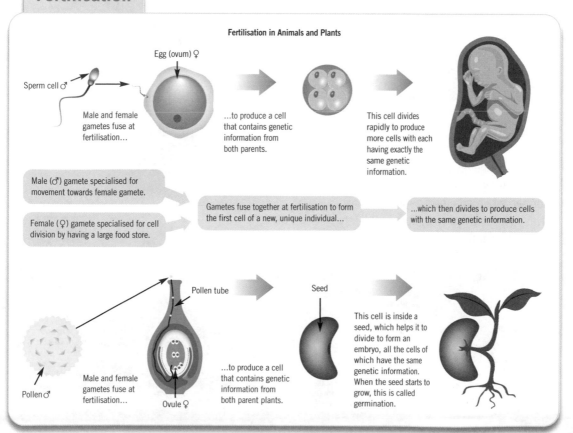

Fertilisation in Animals and Plants

Egg (ovum) ♀

Sperm cell ♂

Male and female gametes fuse at fertilisation...

...to produce a cell that contains genetic information from both parents.

This cell divides rapidly to produce more cells with each having exactly the same genetic information.

Male (♂) gamete specialised for movement towards female gamete.

Female (♀) gamete specialised for cell division by having a large food store.

Gametes fuse together at fertilisation to form the first cell of a new, unique individual...

...which then divides to produce cells with the same genetic information.

Pollen tube

Seed

Pollen ♂

Male and female gametes fuse at fertilisation...

Ovule ♀

...to produce a cell that contains genetic information from both parent plants.

This cell is inside a seed, which helps it to divide to form an embryo, all the cells of which have the same genetic information. When the seed starts to grow, this is called germination.

Variety in the Environment

Asexual Reproduction

Asexual reproduction...
- only needs one organism
- can produce other organisms without the need for genetic material from another individual.

A simple, single-celled animal like an amoeba can divide to produce new genetically identical offspring. Plants can also produce genetically identical offspring, for example, the spider plant.

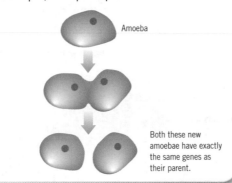

Amoeba

Both these new amoebae have exactly the same genes as their parent.

Cloning

Clones are genetically identical individuals. Identical twins are clones of each other, formed when an embryo splits into two. This is a relatively rare, but completely natural, event.

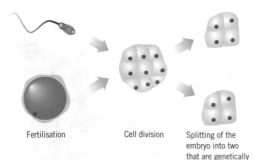

Fertilisation   Cell division   Splitting of the embryo into two that are genetically identical.

Artificial Cloning

In artificial cloning, steps are taken to ensure that all the genetic material comes from one parent:
- In gardening, a simple method is to take **cuttings** of a plant that you'd like to have lots of.

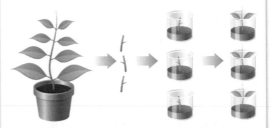

- With animals, the procedure is much more difficult.

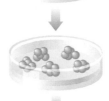

Prize ewe

An egg cell is stimulated to behave as though it's been fertilised.

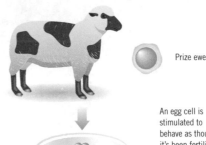

It divides several times to form a ball of cells that is then split apart into several clumps, before the cells become specialised.

These embryos are then implanted into the uteruses of sheep that will eventually give birth to clones.

Competition Between Plants

Plants compete for space, light, water and minerals:

- If plants don't have enough space, they don't have room to grow.
- If they don't have enough light, they can't photosynthesise.
- If they don't have enough water, they will wilt and eventually die.

- If plants don't have enough minerals, they can't perform various chemical reactions, including photosynthesis. Nitrogen, potassium and phosphorus are the main minerals a plant needs. Nitrogen is used to help the plant make protein. Magnesium is also needed to make chlorophyll.

Plants can also be affected by animals and herbicides (chemicals that kill plants).

Use of Fertilisers

Fertilisers...

- can be bought in large bags and spread from the back of a tractor to provide all the necessary minerals a crop needs
- can be washed into rivers if too much is put on the fields.

Weeds

Weeds...

- are plants that aren't welcome in an area
- are 'wild' in the sense that they reproduce as best they can without man's help
- compete vigorously with other plants for space, light, water and minerals.

Use of Herbicides

Selective herbicides...

- can be sprayed onto crops in order to kill particular weeds without harming the food plant the farmer is growing
- remove a vital food resource for many small animals by killing the weeds. This can seriously affect other animals (and possibly the whole food web) by reducing the amount of food available. In some cases, this can result in animals eating more of the food crop to compensate.

In the food web opposite, the removal of weeds to increase growing space for the barley may remove a food source for the dormouse. This may result in the dormouse population having to eat more of the barley crop to survive. In addition, if the dormouse population declines, the fox and owl populations may also decline. This could result in more competition between foxes and hawks for rabbits.

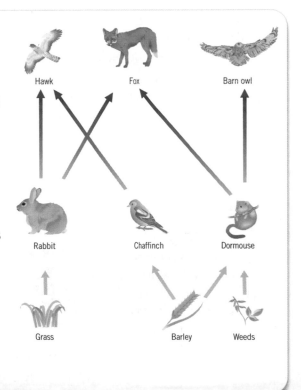

Hawk Fox Barn owl

Rabbit Chaffinch Dormouse

Grass Barley Weeds

Variety in the Environment

Competition Between Animals

Animals compete with each other for...
- food
- **territory**.

Territory is important as each animal needs a space that contains enough food for its survival. A territory also has to support enough animals to allow them to breed successfully.

Pests and Pesticides

Pests of plants...
- include field mice, caterpillars, aphids (greenfly), snails and slugs
- compete with humans for crops, and the more they eat, the fewer there are for humans.

Pesticides...
- are sprayed onto crops to kill animals that eat the crop plants
- improve the appearance of fruit and vegetables, which is extremely important to today's customers. If the food in the shop looks marked, it tends to get left on the shelf.

It's important to remember that each pest is part of a very complicated food web that can be affected by the removal of a certain species.

In addition, the pesticides aren't usually selective and kill useful insects and animals, as well as the pests. So, it might be better (and cheaper) to use a **biological control**, such as a predator insect to eat the pests, rather than a chemical pesticide.

Advantages of Using Pesticides	Disadvantages of Using Pesticides
<ul><li>Increases **crop yield**</li><li>Undamaged produce</li><li>Greater profitability</li></ul>	<ul><li>Effect on the food web</li><li>Accumulation of pesticides in the food web</li></ul>

Quick Test

1. What are gametes?
2. What is the name given to the time when a seed begins to grow?
3. What is a clone?
4. What is a selective herbicide?
5. What is a pesticide?
6. What is a biological control?

KEY WORDS

Make sure you understand these words before moving on!
- Asexual reproduction
- Biological control
- Cloning
- Crop yield
- Cuttings
- Gametes
- Herbicides
- Minerals
- Pesticides
- Selective
- Sexual reproduction
- Territory

Key Words Exercise

Match each key word with its meaning.

Asexual reproduction	Reproductive cells that fuse together at fertilisation
Biological control	Production of new individuals requiring male and female parents
Cloning	Production of new individuals requiring only one parent
Crop yield	Producing exact genetic copies of an organism
Cuttings	Small sections of a plant grown into separate copies in different places
Gametes	The amount of crop produced in a growing season
Herbicides	Using a living organism to keep pest numbers down in a crop
Minerals	Substances that are necessary for chemical reactions such as photosynthesis
Pesticides	Something that only affects certain types of plant or animal, not all of them
Selective	An area needed by animals to allow feeding and breeding
Sexual reproduction	Substances used to kill plants
Territory	Substances used to kill animals such as insects, slugs or other pests

Comprehension

Read the passage about strawberry plants, then answer the following questions.

1. Suggest how it's an advantage for the grower to plant strawberries through plastic sheeting?

2. a) What resources would the runners compete for in the strawberry plant?
   b) Why would this reduce the yield if they weren't removed?

3. The runners are clones of the parent. What does this mean?

4. The fungi are living organisms. What sort of pest control is this an example of?

Strawberries are grown commercially in raised rows in large fields. They're usually planted through holes in plastic sheeting that covers the soil. Irrigation tubes run under the plastic to carry water.

Later in the season, strawberry plants produce runners, as well as flowers for fruit. Runners are stem-like outgrowths along which small plantlets form. For the first two years or so, the runners are cut off because they compete with the flowers for resources in the plant and therefore reduce the yield. After a few years, though, the runners are potted up and allowed to root. Once rooted, the new plants are cut away from the parent plant and used to renew the growing stock.

Chafer grubs are pests of strawberry plants because they eat the roots. Pathogenic fungi are sometimes introduced into the soil when planting the strawberries because they attack and kill the grubs.

Variety in the Environment

Testing Understanding

1 **Fill in the missing words to complete the sentences about sexual and asexual reproduction.**

a) Sexual reproduction depends on the fusion of two _____, which are cells specialised for this job. In humans, they are the _____ and the _____, one of which is specialised for _____ and the other by having a large _____ _____.

b) The moment of fusion is called _____, when the _____ from the two cells combine to form a genetically _____ individual. This new individual now carries _____ from both its _____ and, as a result, may have similar _____ to one or other of them.

c) Asexual reproduction doesn't depend on other individuals at all since only one _____ is involved. Plants are much better at this method of reproduction as only relatively few _____, such as amoebae, are capable of it.

d) The offspring produced by asexual reproduction are _____ and show no _____. Because of this, gardeners often produce more plants by taking _____, as they know the offspring will be exactly like the parent plants.

2 **Read this label from a container of weedkiller called Grassonly, then answer the questions.**

a) What is a selective weedkiller?

b) What do you think is the difference between the shape of grass and weed leaves?

c) Suggest two reasons why you would want to remove weeds from your lawn.

d) If you left large plants like buttercups in your lawn, what would happen to the grass?

e) Why would it be important to wear protective clothing when using Grassonly?

Selective weedkiller, for control of broadleaved weeds in lawns, such as dandelions, clover, buttercups and daisies.
ACTIVE INGREDIENT = 2,4-D Poisonous if swallowed
- Do not spray on windy days.
- Use with caution and prevent spray contact with broadleaved shrubs close to lawn area.
- Wear protective clothing, especially gloves, when mixing and using solution.
- Do not dispose of unused chemical or solution into drains or watercourses.
- Wash hands after use.
- Store in a cool place and keep out of reach of small children.

GRASSONLY

Some pupils set up an experiment in two plots in the school garden to investigate the effect of pesticide on the lettuce crop.

They planted each plot with the same amount of lettuce seeds and gave each plot the same amount of water each day.

After several weeks, when the plants were growing well, one plot was sprayed with a solution of a pesticide, the other sprayed with water.

Each day, the pupils went out and counted the number of caterpillars, slugs and snails that they could find.

Six weeks later, the pupils harvested their lettuces and weighed them. Their results are shown in the tables below.

Plot Sprayed with Water

Week	1	2	3	4	5	6
Number of caterpillars	4	10	14	20	22	8
Number of slugs	2	6	8	12	14	16
Number of snails	0	1	0	3	2	2

Plot Sprayed with Pesticide

Week	1	2	3	4	5	6
Number of caterpillars	0	0	4	10	12	2
Number of slugs	0	0	0	6	9	12
Number of snails	0	0	1	1	2	2

1. Plot the data in the form of two bar graphs, labelling the axes clearly.

2. What variables would the pupils have to assume would remain the same in order to make this investigation a fair test?

3. What is the dependent variable (the variable that's being measured) in this investigation?

4. Does the pesticide affect the pests? Give a reason for your answer.

5. In the end, the water-sprayed plot produced 2450g of lettuce and the pesticide-sprayed plot produced 3020g. Does the pesticide affect the yield? Give a reason for your answer.

6. The numbers of pests went up in both plots almost each week. Why do you think this happened?

7. Suggest a reason why the number of caterpillars dropped in week 6.

Manipulating the Environment

Selective Breeding

Selective breeding is breeding for a desired characteristic. The idea is that new varieties of organisms can be bred by taking advantage of variation:

- Organisms with a desired characteristic are bred with similar organisms.

- This results in offspring, some of which will have an exaggerated version of this characteristic.
- These offspring are then bred again, and so on, until the desired result is achieved.

Selective Breeding in Animals

The diagram below illustrates how new varieties can be produced. Black patches have been selected as the desired characteristic to produce spotty dogs (Dalmatians).

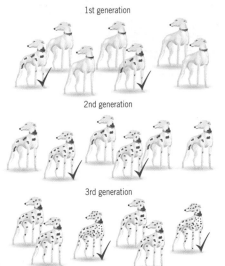

1st generation

2nd generation

3rd generation

In farming, selective breeding is carried out to improve the stock and to develop new varieties of produce. For example, farmers might breed...

- short-legged sheep on upland farms
- cattle that grow to maturity more quickly and produce more beef
- cows that give greater yields of milk
- hens that lay more eggs.

Selective Breeding in Plants

The diagram below shows an example of green vegetables bred from a common ancestor.

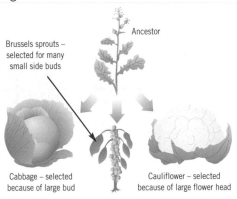

Ancestor

Brussels sprouts – selected for many small side buds

Cabbage – selected because of large bud

Cauliflower – selected because of large flower head

Many well-known fruit and vegetables would be strange-looking things to someone who lived 300 years ago. This is because they've been bred selectively to create products that consumers prefer. However, it takes many generations to get the desired result.

Seedless orange

Wheat with greater seed heads

Dangers of Selective Breeding

Too much selective breeding can cause individual genes to be lost from the breeding population. The loss of a particular characteristic is something to avoid because once it's gone, it's gone forever.

This means there's a loss of variation in the population. Rare breed sanctuaries look after breeds that are no longer used on farms so that useful genes aren't lost forever.

How Human Activities Affect the Environment

Burning fossil fuels, cutting down trees, using pesticides and using fertilisers are examples of how human activity can affect the environment.

Example	Effect on the Environment
Burning fossil fuels and cutting down trees	Fossil fuels (i.e. coal, oil and gas) are being burned at a great rate for electricity and transport. Trees and other plants are also being removed at a great rate for farming or to build homes. These activities contribute to the thickening of the layer of carbon dioxide in the atmosphere and to an increase in the Earth's temperature (global warming). See page 166 for further information.
Pesticides	Pesticides may be washed off the land and into lakes and ponds. Once inside an organism, the pesticide may not break down and may stay in the organism. The food chain below shows how this could affect different organisms: pondweed ➡ tadpole ➡ minnow ➡ perch ➡ heron • A pesticide is taken up by pondweed, which is then eaten by tadpoles. • Each minnow eats a lot of tadpoles, so the amount of pesticide builds up inside it, and each perch eats many minnows. The build-up of chemicals inside an organism is called bioaccumulation. • As a result of bioaccumulation, the levels of pesticide in the herons could be very high. This could weaken the herons' eggshells, making it difficult for the eggs to be incubated without breaking. This might cause a decline in the number of herons.
Fertilisers	If fertilisers run off into water they may cause lots of plants, especially algae, to grow: • At first, this is good because it provides food for animals and supports the food web. • But as the plants die, the microorganisms that decay them use up the oxygen in the water in respiration. This may cause fish to die and could eventually make the entire water stagnant and useless. This process is called eutrophication.

Manipulating the Environment

Energy in Food Chains

Look at the food chain and pyramid of numbers shown opposite:

- The grass uses energy from sunlight to make food by photosynthesis. But most of the light either misses the leaf, is reflected from the leaf or misses the chlorophyll in the leaf, so it isn't used. Actually very little light energy, as a proportion of the total, is used by green plants for photosynthesis.
- The rabbit eats the grass as its energy source. It either uses the energy for respiration, when it's transformed to heat energy and radiated away, or it's not digested and is removed in faeces. In addition, some energy is excreted in urine. So, the rabbit doesn't keep a lot of energy for the fox to use.
- The fox is just like the rabbit in that it loses a lot of energy too. This is why a rabbit has to eat a lot of grass and foxes have to eat a lot of rabbits to obtain enough energy.

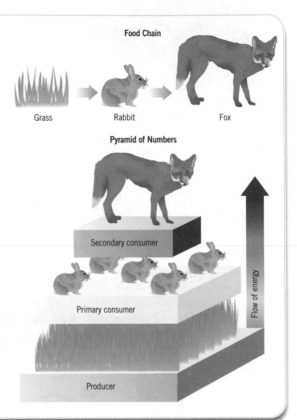

Food Chain

Grass → Rabbit → Fox

Pyramid of Numbers

Secondary consumer

Primary consumer

Producer

Flow of energy

Effects on the Energy in Food Chains

It's very important to check any potential environmental effects before introducing any changes to a food chain. If people are going to manipulate the environment to their needs, they must be sure of the consequences. Here are two examples:

- Limiting factors that reduce the rate of photosynthesis (e.g. the availability of water, light and carbon dioxide, or the amount of chlorophyll) will reduce the amount of energy available in a food chain.
- Changing genes that alter plants to make them outcompete others may affect the energy flow through a system.

Changing Genes – Good or Bad?

Genetic modification...
- is sometimes referred to as genetic engineering
- involves changing genes in organisms, sometimes between different organisms.

For example, a gene from a fish that produces a kind of antifreeze could be added to a tomato to enable it to grow at lower temperatures. The tomato would then be called a GMO (genetically modified organism).

This kind of tomato would cost less to grow and could be grown over a longer period. As a result, the grower could make more profit and it would cost the customer less. But is this a good thing?

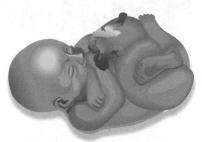

Stem cells...
- come from either embryonic cells (originating from fertilised human eggs) or from the umbilical cords of newborn babies
- can grow into any type of cell.

Consider that stem cells could be grown into brain cells to help sufferers of Parkinson's disease. The cells could be implanted into the patient's brain to replace cells damaged by the disease. Do you think this is a good idea?

A number of questions have to be addressed to reach an answer that meets human needs without damaging the environment (i.e. sustainable development):
- Is it right to genetically modify human cells to try to cure cancers or diseases like cystic fibrosis?
- Should the research into these new technologies be carried out on humans or animals? Would you volunteer to take part in tests?
- What effect might genetic modification have on the environment?

Quick Test

1. What is selective breeding?
2. What is the main danger of selective breeding?
3. What is the main gas contributing to global warming?
4. What gas do the microorganisms of decay use up?
5. Why are food chains usually very short (only three to five steps)?
6. What is a GMO?

KEY WORDS

Make sure you understand these words before moving on!
- Bioaccumulation
- Embryonic
- Eutrophication
- Global warming
- GMO
- Limiting factors
- Selective breeding
- Stem cell
- Sustainable development
- Varieties

Manipulating the Environment

Key Words Exercise

Match each key word with its meaning.

Bioaccumulation	A range of different types within the same species
Embryonic	A cell able to develop into many different types
Eutrophication	Using variation to breed a desired characteristic into the offspring
Global warming	An increase in the Earth's temperature caused by the thickening of the layer of carbon dioxide in the atmosphere
GMO	An increase in the concentration of something along a food chain
Limiting factors	Caused by fertilisers in water, leading to stagnation
Selective breeding	Environmental changes that alter a reaction like photosynthesis
Stem cell	An organism that has had its genetic make-up artificially altered
Sustainable development	Meeting human needs without damaging the environment
Varieties	Originating from the developing fertilised egg

Comprehension

Read the passage about organic farming, then answer the following questions.

1. What does the term 'biodegradable' mean?

2. Why is it important that the compost heap is turned over to allow oxygen into it?

3. How does digging compost back into the soil add nutrients?

4. Light and water are two environmental factors needed for photosynthesis. Why might these be limiting factors if the plants were grown in the wrong places?

5. How might the appearance of cabbages grown organically differ from those grown using pesticides and herbicides?

In organic farming, all biodegradable waste material is composted and dug back into the soil. This means placing all grass clippings, vegetable peelings, old leaves and dead plants in a large heap. This pile is turned over occasionally to make sure oxygen gets to all parts of it. After several months, you have excellent compost that adds nutrients to the soil as well as improving the soil structure.

Plants are only grown where the conditions suit them. If the plant likes shade, then it's not grown in full sunlight. If the plant likes to be well-drained, it's not grown in heavy clay, which doesn't drain away water easily.

No artificial fertilisers, herbicides or pesticides are used. Often slugs, snails and caterpillars are picked off by hand. Netting is used as a physical barrier to flying pests. Weed growth is restricted by growing through plastic sheeting or by mulching the soil with materials like bark chippings.

Testing Understanding

1 **Fill in the missing words to complete the sentences about the effects of pesticides on food webs.**

a) Chemicals are often used to increase crop _____, improve appearance and remove competition. But they also have negative effects on food webs and wildlife.

b) Pesticides, a general name for fungicides, insecticides and rodenticides, are, as their name suggests, used to combat the problem of _____. They have improved the appearance of our fruit and _____, which are no longer ravaged by pests such as caterpillars, _____ (aphids) or field mice.

c) However, killing populations of pests drastically affects _____ webs. For example, imagine that thrushes eat insects and slugs; if all the _____ were killed by insecticides, the thrushes' _____ source would be severely _____ and thus the numbers of thrushes would also be reduced. The consequence of this is that _____ would have _____ predators and may, therefore, flourish. They will in turn eat _____ of the food crop.

d) _____ can also be washed into ponds and lakes, where they are broken up by microscopic plants. At this stage they're very _____, but become more and more _____ at every stage of the food chain. Eventually they become concentrated enough to _____. This is called _____.

2 **Read the information about the effect of DDT on peregrines, then answer the questions that follow.**

Extensive use of the insecticide DDT, to kill insects, during the 1950s resulted in a marked decline in the number of peregrine falcons. DDT can't be excreted by animals that eat it and the accumulation of the poison in the falcons caused them to produce very thin eggshells that often cracked, killing the chick.

The following data refers to the number of sightings of peregrines after the use of such pesticides was banned:

Year	1964	1966	1968	1970	1972	1974	1976	1978	1980	1982	1984	1986
No. of Sightings	4	5	4	6	7	6	12	11	15	22	43	76

a) Plot this data in the form of a bar graph.
b) Suggest how many sightings there may have been in 1985.
c) In which year would the number of sightings have been approximately 33?
d) How many years did it take for the peregrine population to treble?
e) Peregrines eat mainly small mammals, so why did an insecticide nearly wipe them out?

Manipulating the Environment

Hardeep and Sue were doing an experiment in plots in the school garden. They planted carrots into six different plots.

At the start of the experiment, immediately before planting, they added an artificial fertiliser, Vegegrow, to the plots in different amounts.

Each plot was treated in the same way every day. After several months, they carefully dug up the carrots, and washed and dried them before weighing them.

Their results are shown in the table below.

Plot	1	2	3	4	5	6
Amount of Fertiliser (g)	0	4	8	12	16	20
Average Mass of Carrot (g)	262	322	368	402	404	398

1. Draw a bar graph to show how the average mass of carrot varied from plot to plot.

2. Why didn't adding more fertiliser make much difference in plots 5 and 6?

3. What are the environmental dangers of adding too much fertiliser?

4. What is the dependent variable in this investigation?

5. What variables would need to be controlled in order to make this a fair test?

Social Interaction

Observing Behaviour

Behaviour is the organism's response to changes in its internal or external environment. So, if you wish to observe responses, you can investigate them by manipulating the environment and recording the responses.

The table below shows some examples.

Observation Test	Description
Watching woodlice	If a number of woodlice are placed in the centre of a large Petri dish, they will run around very fast in all directions. When they reach the edge of the dish they will slow down and move around the edge, often stopping when they meet another woodlouse. This behaviour can be watched and recorded on film, or the positions of the woodlice plotted every minute or so for a period of time.
Watching maggots	If a narrow beam of light is shone onto a fly maggot's head, it will move away from it. So, you can shine a light on the maggot from one direction and record the response, then repeat it from another direction and so on. If the maggot is covered with a dilute, coloured liquid (that doesn't affect it!), then the trail it leaves behind can be analysed in relation to the light directions.
Watching humans	Humans respond to exercise with increases in heart rate, breathing rate and skin temperature. These responses can be measured before and after vigorous exercise and recorded.

Social Interaction

Benefits of Behaviour

In each example on the previous page, the organisms benefit from the behaviour they show:

- Woodlice – the random running around enables them to find shelter from predators as quickly as possible. Being in touch with an object keeps them protected.
- Maggots – turning away from light keeps them inside the food material (for example, rotting meat) and away from predators.
- Humans – increased breathing and heart rates provide more oxygen to the body and move the blood faster, which means energy gets to the muscles more quickly. The skin temperature rises to cool the blood so that the internal body temperature doesn't rise too high.

Here are some other examples of beneficial behaviour:

- A baby will cry when its nappy is wet to draw attention to itself, so that it can be changed. This prevents the baby becoming uncomfortable and reduces the chances of nappy rash.
- Baby birds will call their parents for food to allow them to grow and develop.
- Parent animals will make warning noises to their young to tell them that they're in danger, as a predator may be around.

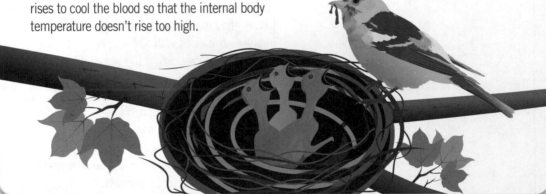

How Plants Behave

Plants don't really show behaviour as such, but they do respond to **stimuli**:

- When a seed germinates (starts to grow), the root always grows down and the shoot always grows up as a response to gravity.
- Young seedlings will grow towards the light when placed on a windowsill.

This behaviour ensures that the leaves are in the best possible light and the roots are in the soil to obtain nutrients.

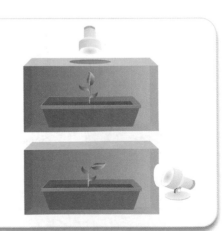

Learning Behaviour

Behaviour is often learned from interaction with other individuals, for example, parents, brothers and sisters, and peers:

- Animals may learn from their parents which foods are good to eat and may learn how to hunt. Play also helps young animals to practise the skills required to survive.
- Within any social group, behaviour is learned to retain the position within it. Gorillas will groom each other to maintain their group position. Male animals will learn how to show that another is superior and therefore will not come to harm.
- **Courtship displays**, including song, often 'mark' territory for a male as well as attract females to it. These behaviours are often learned through experience.
- People learn traditional customs that last for many years to help maintain the community and social life in which they live. You also develop **habits**, which are often very hard to break, through a series of short, learned steps.

In all cases, those organisms that show the most suitable behaviour are the most likely to succeed in terms of producing offspring. This is of huge benefit to the **species** (similar organisms that can **mate** to produce fertile offspring).

Living Together

All animal **populations** (groups of the same species) rely on the behaviour of the individuals in it to survive.

Even solitary animals have to use behavioural responses when they meet another member of the species, or when they need to mate.

Behaviour for Mating

Here are some examples of how different organisms behave at mating time:

- Corals use a variety of environmental signals (e.g. temperature and the cycle of the Moon) to synchronise the release of sperm and eggs into the water. By releasing these at the same time, the corals ensure that the eggs have the best chance of **fertilisation**.
- Male crickets sing when they're ready to mate. The 'song' changes when a female approaches. They touch antennae and mating occurs. The singing increases the chances of males and females meeting, which makes fertilisation more likely.
- Mute swans mate for life, which helps to ensure that eggs are fertilised each year. To maintain the relationship, the pair show courtship display behaviour at mating time.

Social Interaction

Behaviour for Status

Here are some examples of how different social behaviours help individuals and populations to exist side by side in relative comfort and allow successful reproduction:

- When you meet royalty you bow, at least your head, as a means of showing that they have higher status in the population.
- When people meet each other, they often greet each other or shake hands to acknowledge that they don't threaten each other.

- When an inferior lion meets a bigger, stronger lion, it will show respect by backing away and placing its tail between its legs. This prevents fighting and injury to either lion.

This kind of social behaviour is necessary for different species to survive.

Quick Test

1. What is the total response of an organism to changes in the environment called?
2. What is the name given to the factor that's sensed by an organism?
3. What is the benefit to a maggot of turning away from light?
4. What type of behaviour is the display of a male to a female?
5. What is the name given to the interaction behaviour between all members of a group?

KEY WORDS

Make sure you understand these words before moving on!

- Behaviour
- Courtship display
- Fertilisation
- Habit
- Mating
- Population
- Response
- Species
- Status
- Stimulus

Key Words Exercise

Match each key word with its meaning.

Behaviour	Something that causes a response in the nervous system
Courtship display	The action made due to receiving a stimulus
Fertilisation	The pattern of responses of individuals to a stimulus or stimuli
Habit	The meeting of a male and female for reproduction
Mating	Learned behaviour gained by a series of steps
Population	Behaviour designed to allow reproduction to occur
Response	The total number of individuals of the same species in an area
Species	An individual's position within a group
Status	Similar organisms able to breed to produce fertile offspring
Stimulus	The fusing of male and female nuclei

Comprehension

Read the passage about adult red deer, then answer the following questions.

1. What is meant by the 'rut'?

2. a) What is the purpose of the rival stags walking alongside each other?
   b) What is the advantage of this walk and possible fighting?

3. What is the purpose of 'roaring'?

4. Why do you think the hind keeps her offspring with her for one year?

Adult red deer are found in single-sex groups for most of the year. During the mating season, from late August to early winter, called the rut, mature male stags compete for the female hinds. They will challenge each other by walking alongside each other and assessing each other's size, antlers and likely fighting ability. If neither backs down, fighting may break out in which the stags use their antlers to force the other to retreat.

The stronger stags will attract groups of hinds by continuous 'roaring' and this also keeps his harem (the hinds) near to him. The stag will mate with each female in the harem. Each female produces one calf (very occasionally two) after about 8–9 months and keeps it with her for a further year.

Social Interaction

1 **Fill in the missing words to complete the sentences about social interaction.**

  a) The total responses of an individual to its environment are called its _____.

   This allows the individual to adapt to any _____ in the internal or

   external environment.

  b) You can observe _____ by artificially changing the environment. The purpose of

   this behaviour is to survive. It is often to _____ against predators or to

   keep the individual in the most favourable environment.

  c) The parent animal, or interaction with other individuals, helps the young to

   _____ the appropriate behaviour. This helps the youngster to keep its

   place, or _____ in the group, and also allows it to learn how to attract a

   member of the opposite sex for _____.

  d) The overall _____ behaviour has a beneficial effect on the survival of

   the species.

2 **Read the information below, then answer the questions that follow.**

Male zebra finches show courtship behaviour towards females. An experiment was carried out
in which male finches were caged with...
* a female with a red beak (cage A)
* a female with a black beak (cage B)
* a model female with a grey beak (cage C).

These were the results:
* In cage A, 88% of males showed courtship behaviour for an average of 7 seconds.
* In cage B, 64% of males showed courtship behaviour for an average of 4 seconds.
* In cage C, 42% of males showed courtship behaviour for an average of 1 second.

  a) Plot this data in the form of two separate bar graphs, with the females shown on the x-axis.
  b) What conclusion can you make about the number of responses by the males?
  c) What conclusion can you make about the length of the courtship behaviour shown by the males?
  d) What overall conclusion can you make from this data?
  e) What was the independent variable in this experiment?

Shannon and Jason carried out an investigation into the behaviour of woodlice in different humidity levels (the amount of water vapour in the air).

Woodlice are crustaceans, like crabs and shrimps, and breathe through gills, not lungs.

Shannon and Jason set up six Petri dishes, each containing a different and known percentage relative humidity.

A single woodlouse was placed into each dish and the distance it moved in a set period of time was recorded.

This was repeated several times and the average results are given in the table below.

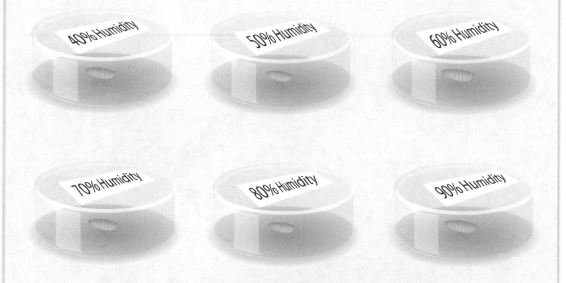

Relative Humidity (%)	40	50	60	70	80	90
Centimetres Moved per Minute	22	21	21	18	12	10

1 Plot the data in the form of a line graph (label the axes clearly).

2 What variables would Shannon and Jason have to assume would remain the same in order to make this investigation a fair test?

3 What is the dependent variable in this investigation?

4 What conditions should the woodlice have been kept in before the experiment?

5 a) Why should only one woodlouse at a time be placed in the Petri dish?
   b) Why did Shannon and Jason repeat this several times?

6 How would the response of woodlice to humidity help them to survive in natural conditions?

Chemistry

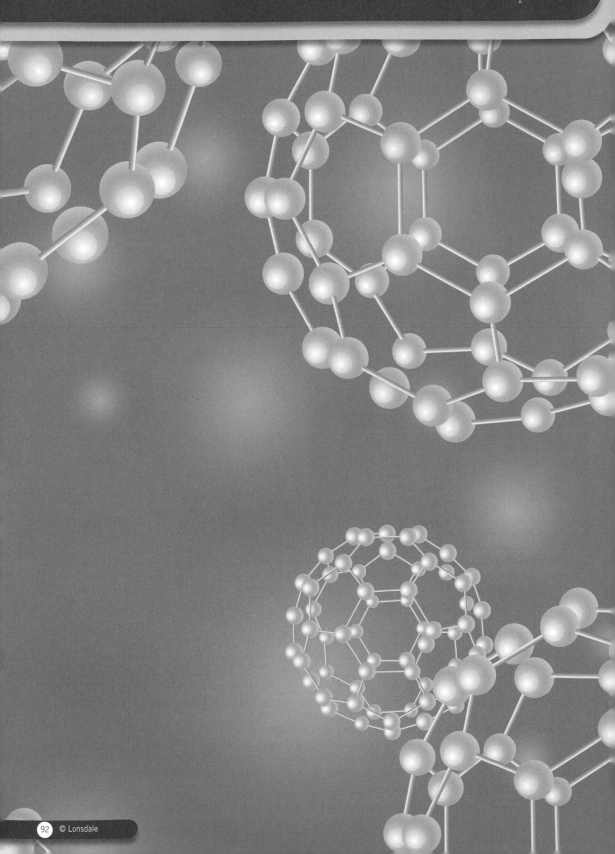

Acids and Alkalis

Acids

Acids...
- have a **pH** of less than 7
- react with alkalis to form **neutral** solutions.

Weak acids are safe to handle. They can be found in certain foods, e.g. vinegar and lemon juice. This is what gives the food a **sour** taste.

Some foods are preserved by pickling them in vinegar. The acid in the vinegar stops the bacteria (which can make food go off) from multiplying.

Strong acids for example, sulfuric acid (found in car batteries) and hydrochloric acid, can be **corrosive** (wear things away). Special care must be taken when using them.

Acids are used to produce many useful chemicals, for example...
- fertilisers
- plastics
- explosives.

Alkalis

Alkalis...
- have a pH greater than 7
- react with acids to form neutral solutions.

Weak alkalis can be handled safely. They can be found in products like toothpaste and shower gel.

Strong alkalis, are hazardous and they must be used very carefully. Some cleaning materials, for example, drain cleaner, contain strong alkalis.

Alkalis are used to produce many useful materials, for example...
- paper
- glass
- soap.

Acids and Alkalis

Safety Rules

Make sure you follow these rules when dealing with acids and alkalis:

- Never taste any chemicals in the laboratory.
- Always wear goggles.
- If you spill some acid or alkali on your skin, wash it off using lots of water, and tell your teacher.
- If you spill a large amount of acid or alkali, tell your teacher straight away. They will sprinkle sand onto the chemical, so it can then be swept up and disposed of safely.

Hazardous Substances

Although many household acids and alkalis are harmless and can be handled quite safely, other acids and alkalis are **hazardous**.

Hazard symbols are...

- displayed on the labels of containers
- used to warn people about the dangers of the chemicals they're using.

You should be able to recognise the hazard symbols in the table opposite.

When chemicals are transported by road, the vehicles carrying the chemicals must display the appropriate hazard-warning signs.

If the vehicle is involved in an accident, the signs tell the emergency service workers how to deal with the chemicals carried by the vehicle.

If large amounts of acid are released in an accident, the emergency service workers add water to the acid.

Adding water **dilutes** the acid and makes it less hazardous.

Hazard Symbol	Explanation
![X h]	Harmful substances – similar to toxic substances, but less dangerous.
![corrosive]	Corrosive substances – attack and destroy living tissues, including eyes and skin.
![X i]	Irritants – not corrosive but can cause reddening or blistering of the skin.

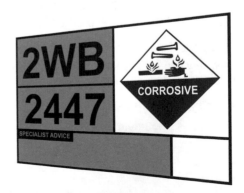

2WB
2447
SPECIALIST ADVICE
CORROSIVE

Neutral Solutions

- Neutral solutions have a pH of 7.
- They can be made by reacting an acid with an alkali.
- Water and alcohol are neutral solutions.
- Some neutral solutions are hazardous, for example, alcohol is highly flammable.

Neutralisation

- Acids and alkalis are chemical opposites.
- If an acid is added to an alkali the pH of the alkali goes down.
- If exactly the right amount of acid is added to an alkali then the acid **neutralises** (cancels out) the alkali and a neutral solution is formed, i.e. the solution will have a pH of 7.

Using Neutralisation Reactions

Neutralisation reactions can be used to change the pH of certain substances.

Use	Explanation
Soil treatments	Many plants only grow well in soil that has a certain pH: • Lime (calcium hydroxide, a base) can be added to soil to make it more alkaline. • Acidic compounds can be added to soil to make it more acidic.
Indigestion treatments	• Antacid tablets cure indigestion by neutralising the extra hydrochloric acid produced by the stomach.
Skin care	• Some shower gels and moisturisers are pH balanced so they have a similar pH to your skin.
Hair care	• Shampoos contain alkalis that clean hair by removing grease. • Conditioners contain acids that neutralise the alkalis in the shampoo.
Insect bites	• Vinegar (an acid) can be used to treat wasp stings (alkaline). • Camomile (an alkali) can be used to treat bee stings (acidic).

Acids and Alkalis

Indicators

Indicators contain a dye that's one colour when it's mixed with an acid and another colour when it's mixed with an alkali. Some plants, such as red cabbage and beetroot, contain dyes that can be used as indicators.

Universal indicator contains a mixture of different dyes.

It can be used to tell how acidic or how alkaline a solution is.

pH Scale

The strength of an acid or an alkali can be measured using the pH scale:

- Neutral solutions have a pH of 7.
- Acids have a pH less than 7.
- Alkalis have a pH greater than 7.
- Strong acids are much more corrosive than weak acids.

- Strong alkalis are much more corrosive than weak alkalis, and are sometimes described as being caustic.

The colour chart below shows the colour that universal indicator turns when it's mixed with solutions that have different pHs.

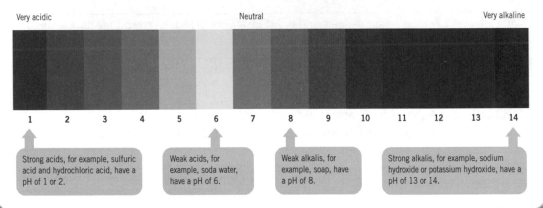

Very acidic   Neutral   Very alkaline

1   2   3   4   5   6   7   8   9   10   11   12   13   14

Strong acids, for example, sulfuric acid and hydrochloric acid, have a pH of 1 or 2.

Weak acids, for example, soda water, have a pH of 6.

Weak alkalis, for example, soap, have a pH of 8.

Strong alkalis, for example, sodium hydroxide or potassium hydroxide, have a pH of 13 or 14.

Quick Test

1. What type of substance has a pH greater than 7?
2. What are hazard symbols used for?
3. Give an example of a substance with a neutral pH.
4. What is the pH scale used to measure?

KEY WORDS

Make sure you understand these words before moving on!

- Acid
- Alkali
- Corrosive
- Dilute
- Goggles
- Hazard symbol
- Hazardous
- Indicator
- Lime
- Neutral
- Neutralise
- pH
- Property
- Sour
- Universal indicator

Key Words Exercise

Match each key word with its meaning.

Acid	A chemical that has a pH of less than 7
Alkali	A chemical that has a pH of more than 7
Corrosive	To cancel out
Dilute	A scale used to measure how acidic or alkaline a solution is
Hazard symbol	A chemical that has a pH of 7 and is neither acidic nor alkaline
Hazardous	A sharp, acidic taste
Indicator	A substance that attacks and wears away living materials, metals and rocks
Lime	Something that might cause an accident to happen
Neutral	A symbol used to warn people about the chemicals in a container
Neutralise	An alkaline compound used to neutralise acidic soils
pH	To water down
Sour	A chemical that's one colour when mixed with an acid and another colour when mixed with an alkali
Universal indicator	A mixture of indicators that can be used to test how acidic or alkaline a solution is

Comprehension

Read the passage about stinging nettles and dock weeds and then answer the following questions.

1. Where can nettles be found growing?

2. What sort of soil do nettles and dock weeds grow well in?

3. How can someone get a nettle sting?

4. Why are nettle stings so painful?

5. Describe a dock weed leaf.

6. How can nettle stings be treated?

7. Why do dock weed leaves relieve the pain of nettle stings?

Stinging nettles and docks are common weeds. They are often found growing on roadsides and on areas of rough ground. They are found close to each other because they both thrive in nitrogen-rich soil.

The leaves of stinging nettles are covered in delicate hollow hairs. If someone touches these leaves the hairs can be broken, releasing chemicals onto the person's skin and causing a sting. Nettle stings are painful because they contain acid.

Dock weeds have large, dark green oval-shaped leaves. The leaves of dock weeds are a traditional remedy for treating nettle stings. When dock weed leaves are crushed they release a sap, which can be rubbed on to the sting.

Dock leaves contain an alkali, which can neutralise the acid on the skin and reduce the pain caused by the sting.

Acids and Alkalis

Testing Understanding

1 Fill in the missing words to complete the sentences about acids and alkalis.

a) Acids and _____ are chemical opposites.

b) Acids have a pH _____ than 7. Some foods like lemons and vinegar

contain _____ acids. These foods have a sharp, _____

taste. Strong acids like sulfuric acid and hydrochloric acid are _____ and

must be handled with care.

c) Alkalis have a pH _____ than 7. _____ alkalis like soap

can be handled quite safely.

d) Neutral solutions like water and alcohol have a pH of _____.

e) Indicators are special chemicals that are used to tell whether a solution is acidic or

_____. When an _____ is mixed with an acid it turns one

colour. When it's mixed with an _____ solution it turns a different colour.

f) Universal indicator turns a _____ colour when it's mixed with a strong

acid. It's green when mixed with a _____ solution and

_____ when mixed with a strong alkali.

2 Read the information, then answer the questions that follow.

Many metals react with acids. One of the products of the reaction is hydrogen gas.

Britney designed an experiment to compare the reactivity of different metals by counting the number of bubbles of hydrogen produced as each metal was placed in acid. Her results are shown in the table opposite.

a) Draw a bar chart to show these results.
b) Use your graph to place the metals in order from the most reactive to the least reactive.

Metal	Number of Bubbles Produced in 1 minute
Copper	0
Magnesium	21
Iron	10
Zinc	14
Lead	2

Hydrochloric acid is produced in the stomach. It helps us to digest food, but if too much hydrochloric acid is produced it can cause indigestion.

Angela wanted to find out which type of antacid (A, B, C or D) would increase the pH of hydrochloric acid the most.

She placed 50cm³ hydrochloric acid into four separate beakers. She used a pH meter to measure the pH of the acid in each beaker. The pH of the acid was 1.

Angela placed one antacid tablet into each beaker and let a reaction take place. When the reaction stopped, she measured the pH of the solution in each beaker. These are her results:

- Beaker A had a pH of 5.
- Beaker B had a pH of 7.
- Beaker C had a pH of 6.
- Beaker D had a pH of 4.

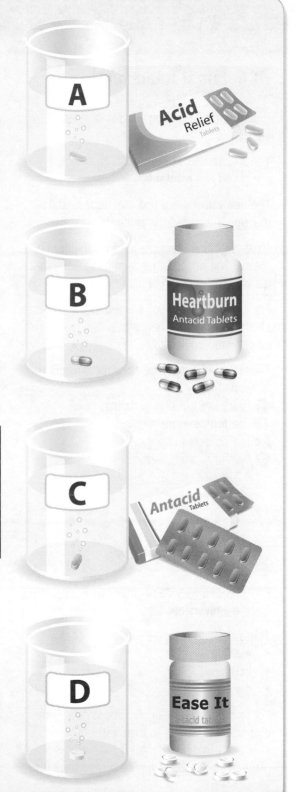

❶ Copy this table and complete it using Angela's results.

Antacid	pH
A	

❷ Draw a bar chart to display the results from Angela's experiment.

❸ What is the independent variable in this experiment?

❹ What is the dependent variable?

❺ How did Angela make this a fair test?

❻ Name the piece of equipment that Angela used to measure the amount of acid.

❼ Which is the most effective antacid?

Chemical Reactions

Chemical Reactions

In a **chemical reaction** new substances are made.

The substances at the start of the reaction are called the **reactants**.

The new substances that are made by the reaction are called the **products**.

Most chemical reactions are **irreversible**. This means that once the reaction has taken place it's very difficult to get the reactants back.

For example, when you place bread in a toaster chemical reactions take place. The toast can't be changed back into bread.

These are all signs that a chemical reaction is happening:
1. There are bubbles or **fizzing.**
2. The **temperature** is changing.
3. The colour is changing.
4. A solid is being formed.

Physical Changes

Physical changes do not make new substances. They are **reversible**.

These are all examples of physical changes:
- **melting**
- **boiling**
- dissolving.

For example, when an ice cube is taken out of a freezer it melts to form water (a liquid). This is a physical change. If the water is placed in a freezer it can be turned back into ice (a solid).

Reacting Acids with Metals

When a metal is reacted with **hydrochloric acid**, you can tell that a chemical reaction is taking place because you can...

- see bubbles – this tells you that a gas is being made
- feel a temperature increase
- see the piece of metal getting smaller. (This happens because acids are **corrosive** and they wear away the metal.)

The reaction stops when either all the metal or all the acid is used up. You can tell that the reaction is over when no more bubbles are being made.

The gas that's made by the reaction burns with a squeaky 'pop'. This tells you that the gas is hydrogen.

Reactions between metals and acids are irreversible.

When hydrochloric acid is added to...

- fairly reactive metals, e.g. magnesium and zinc, a chemical reaction takes place
- very unreactive metals, e.g. copper and gold, no reaction takes place.

Some very reactive metals, for example, sodium and potassium, shouldn't be placed in acids because they would react too quickly.

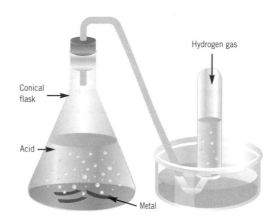

Hydrogen gas

Conical flask

Acid

Metal

Reacting Acids with Metal Carbonates

Calcium carbonate is found in rocks, for example, chalk, marble and limestone. These rocks are sometimes used to make buildings or statues.

You can test whether a rock contains calcium carbonate by dropping hydrochloric acid on to it. If the rock contains calcium carbonate then it starts to fizz. This shows you that a gas is being made.

The calcium carbonate will get smaller as it is corroded (worn away) by the acid. In this way, acid rain attacks and damages buildings and statues that are made from rock containing calcium carbonate.

The reaction stops when either all the calcium carbonate or all the hydrochloric acid is used up. Other metal carbonates, for example, sodium carbonate, react in a similar way.

If the gas that's made by this reaction is bubbled through limewater it turns the **limewater** cloudy. This shows that the gas is carbon dioxide.

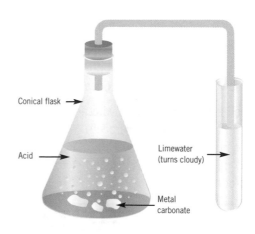

Conical flask

Acid

Limewater (turns cloudy)

Metal carbonate

Chemical Reactions

Burning Substances

Oxygen makes up about 20% of the air. When substances are burned in air they react with oxygen to form new substances called **oxides**. These reactions are irreversible.

When magnesium (a shiny metal) is burned in air, it reacts **vigorously** and burns with a brilliant white light. The magnesium reacts with the oxygen to form magnesium oxide (a white powder).

When copper (a metal) is burned in air it reacts slowly; it glows red and then turns black. The new substance formed is called copper oxide.

When magnesium is burned in pure oxygen, the reaction is very vigorous and special care must be taken.

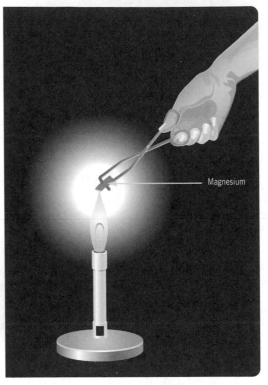

Magnesium

Burning Fuels

Fuels such as coal, oil and natural gas are substances that can be burned to release energy.

When fuels burn they react with oxygen to form new substances called oxides.

Coal, oil and natural gas are hydrocarbons; they contain carbon and hydrogen. When they're burned...
* the carbon reacts with oxygen to form carbon dioxide

* the hydrogen reacts with oxygen to form water vapour.

Many fuels also contain small amounts of sulfur. When sulfur is burned it reacts with oxygen to form sulfur dioxide.

Products of Combustion

You can use the equipment below to find the products formed when a hydrocarbon fuel, for example, methane, is burned.

The hydrogen in methane reacts with oxygen to form water vapour, which **condenses** as it touches the cold surfaces of the glass U-tube.

You can tell that water is present by testing the colourless liquid with **cobalt chloride** paper. It turns from blue to pink in water.

The carbon in the fuel reacts with oxygen to form carbon dioxide. The carbon dioxide turns the limewater cloudy.

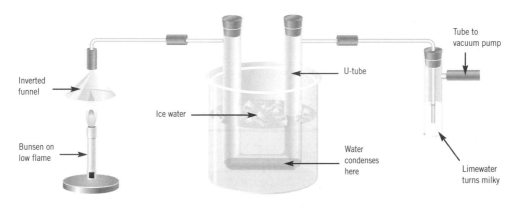

Inverted funnel

Bunsen on low flame

Ice water

U-tube

Tube to vacuum pump

Water condenses here

Limewater turns milky

Fire Safety

The fire triangle shows the three substances that are required for a fire:
- fuel
- heat
- oxygen.

If any one of these substances is removed then the fire will go out.

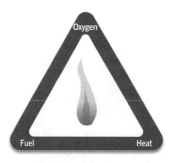

Oxygen

Fuel

Heat

Quick Test

1. Are most chemical reactions reversible or irreversible?
2. What happens to limewater when carbon dioxide is bubbled through it?
3. Copy and complete this word equation: magnesium + oxygen → _____
4. What three things are needed in order for a fire to burn?

KEY WORDS
Make sure you understand these words before moving on!
- Boiling
- Calcium carbonate
- Chemical reaction
- Cobalt chloride
- Condense
- Corrosive
- Fizzing
- Fuel
- Hydrochloric acid
- Irreversible
- Limewater
- Melting
- Oxide
- Product
- Reactant
- Reversible
- Temperature
- Vigorously

Chemical Reactions

Key Words Exercise

Match each key word with its meaning.

Chemical reaction	A change that makes new substances
Condense	A chemical used to test for the presence of carbon dioxide
Corrosive	Compound made when a substance reacts with oxygen
Fuels	The chemicals at the start of a reaction
Irreversible	The chemicals at the end of a reaction
Limewater	A change that can't be undone
Melting	A change that can be undone
Oxide	A change of state from solid to liquid
Products	Substances that can be burned to release energy
Reactants	A change of state from gas to liquid
Reversible	A substance that attacks and wears away living materials, metals and rocks

Comprehension

Read the passage about fireworks and then answer the following questions.

1. Suggest some occasions when fireworks are used.

2. What is a fuel?

3. a) What is an oxidising agent?
   b) Why are oxidising agents used in fireworks?

4. Why do many fireworks contain metal salts?

5. Why must fireworks be treated with great care?

Fireworks contain a mixture of different chemicals, including a fuel (such as carbon), which releases energy when it's burned.

Fireworks also contain an oxidising agent, which is a chemical that releases oxygen. The oxygen released helps the fuel to burn even better.

Many fireworks also contain metal salts, which release bursts of coloured light when they are heated. Different metal salts release different coloured light, for example, copper salts give off blue-coloured light.

Firework displays are used to celebrate special occasions such as Bonfire night, Diwali and New Year's Eve. Lots of people enjoy watching firework displays every year, but it's important to remember that fireworks are explosives and they must only be used by responsible adults.

Testing Understanding

1 **Fill in the missing words to complete the sentences about chemical reactions.**

a) The chemicals at the start of a reaction are called the _____ . The

chemicals made by the reaction are called the _____ . Most chemical

reactions are irreversible.

b) Physical changes like melting and dissolving don't make any new substances. They

are _____ .

c) When substances are burned they react with _____ in the air to form

new substances called oxides. When magnesium is burned it produces magnesium

_____ . When copper is burned it produces _____ oxide.

d) When carbon is burned in a good supply of oxygen, _____ dioxide is made.

e) When hydrogen is burned it reacts with _____ to form water vapour.

f) When sulfur is burned it reacts with oxygen to form _____ dioxide.

2 **Read the information about the experiment with metals and acids, then answer the questions that follow.**

Ed carried out an experiment to find out which metals react with hydrochloric acid.

He poured hydrochloric acid into four test tubes, then took the temperature of the acid and recorded it in a table. He then placed a piece of metal into one of the test tubes of acid, measured the new temperature and recorded this in the table.

He repeated his experiment using the other metals. This table shows Ed's results:

Metal	Temperature of Acid at Start (°C)	Temperature of Acid at End (°C)	Temperature Change (°C)
Zinc	20	35	
Iron	20	28	
Magnesium	21	47	
Copper	20	20	

a) Copy and complete the table to show the rise in temperature.
b) Draw a bar chart to show the temperature change.
c) Which metal didn't react with the acid?

Chemical Reactions

Skills Practice

Cameron wanted to find out if the volume of water in a beaker affected how quickly it warmed up.

He placed 25cm³ water in the first beaker, 50cm³ water in the second beaker, 75cm³ water in the third beaker and 100cm³ water in the fourth beaker.

The temperature of the water in each beaker was 20°C at the start of the experiment.

He heated each beaker for exactly 2 minutes. Then he turned off the heater, stirred the water and measured the temperature of the water.

Here are Cameron's results:
- The temperature of the water in the beaker with 50cm³ water was 60°C.
- The temperature of the water in the beaker with 25cm³ water was 75°C.
- The temperature of the water in the beaker with 75cm³ water was 45°C.
- The temperature of the water in the beaker with 100cm³ water was 30°C.

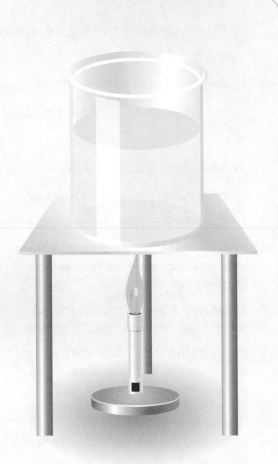

❶ Copy and complete the table opposite using Cameron's results.

❷ What is the independent variable?

❸ What is the dependent variable?

❹ Give one way in which Cameron made this a fair test.

❺ What piece of equipment did Cameron use to measure the temperature of the water?

❻ What effect does the volume of water have on the temperature of the water after it has been heated for two minutes?

Volume of Water (cm³)	Temperature of water after 2 minutes (°C)
25	
50	
75	
100	

For further practice, go to p.20-23 of the Year 7 Science Essentials Workbook

Particles

Particles and Matter

One of the oldest ideas in science is that everything is made up of tiny pieces of **matter** called **particles**.

Matter can be divided into three groups:

- solids
- liquids
- gases.

The way in which the particles are arranged and move is called **particle theory**.

Particle theory explains why solids, liquids and gases behave in different ways.

Solids

Solids, for example, bricks, have a fixed shape and **volume** (size) because the particles have fixed places and are held together by strong forces of **attraction**. In solids, the particles don't move about; they can only **vibrate**.

Solids are **incompressible** (very hard to squash) because the particles are already close together so they can't be pushed any closer.

Solids have a high **density** (heavy for their size) because the particles are close together.

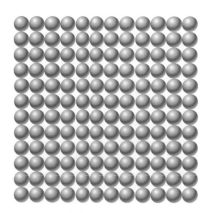

Liquids

Liquids, for example, water, have a fixed volume but they don't have a fixed shape.

This means liquids can **flow**, because the particles in a liquid are able to move past each other. This is because the forces of attraction between the particles are weaker than in solids. So, if a liquid is poured into a container, it takes the shape of the container.

Liquids are dense and incompressible because the particles are close together.

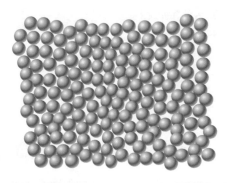

Particles

Gases

Gases, for example, oxygen, don't have a fixed shape or a fixed volume. This means that they will quickly fill any container into which they are placed.

This is because there are no forces of attraction between the particles. The particles move very quickly in all directions.

Gases have a low density (are light for their size) and are **compressible** (easy to squash) because there's a lot of space between particles.

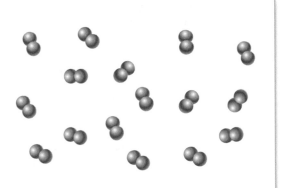

Identifying Solids, Liquids and Gases

You can use a key to identify whether a material is a solid, liquid or gas.

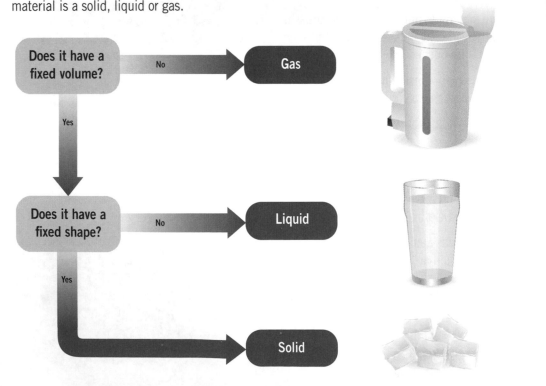

Does it have a fixed volume? — No → Gas

Yes ↓

Does it have a fixed shape? — No → Liquid

Yes ↓

Solid

Expansion and Contraction

When a solid is heated it **expands** (gets bigger) and when it is cooled down it **contracts** (gets smaller).

The following experiment shows how a metal bar changes as it's heated:

Stage of Experiment	Gauge Test	Effect on the Particles
When the metal bar is cold, the particles are vibrating slowly. So, the cold metal bar can fit into the gauge.		
As the metal bar is heated in a Bunsen flame, the particles in the bar vibrate more and the metal bar expands. The hot metal bar no longer fits into the gauge because the bar has expanded.		
When the metal bar cools down, the particles vibrate less and the bar contracts. The cold metal bar now fits into the gauge again, because the bar has contracted.		

Diffusion

Perfumes have quite low boiling points. At room temperature some particles **evaporate** (change from liquid to gas) from the surface of the perfume and mix with air particles. The gas particles move quickly in all directions and quickly spread though the whole room.

Diffusion also takes place in liquids, but because the particles in liquids move more slowly, it takes a little longer. For example, if a purple crystal is placed in a beaker, the crystal dissolves and then slowly mixes with the water particles.

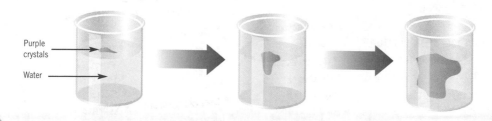

Purple crystals

Water

Particles

Gas Pressure

The gas particles in a balloon move very quickly in all directions. As the particles **collide** with (hit) the inside of the balloon they create gas **pressure**.

The higher the temperature...
- the faster the particles move
- the harder and more often the particles collide with the inside of the balloon
- the higher the gas pressure.

Atmospheric Pressure

The gas particles in the air around you are continually colliding into you and exerting a pressure called atmospheric pressure, which is very strong.

For example, 'empty' plastic bottles are in fact full of air. The pressure inside the bottle is caused by the air particles colliding with the inside of the bottle. The pressure outside the bottle is caused by atmospheric pressure, as the air particles collide with the outside of the bottle. The pressure inside and outside the bottle cancel each other out.

But, if the air particles inside the bottle are removed, the pressure outside the bottle is suddenly greater than the pressure inside the bottle and the bottle dramatically collapses due to the atmospheric pressure.

Air sucked out by vacuum pump

Mineral Water

Quick Test

1. Which is more dense: a liquid or a gas?
2. By what method does a perfume smell spread across a room?
3. What happens to the pressure of a gas as the temperature gets higher?

KEY WORDS
Make sure you understand these words before moving on!
- Attraction
- Collide
- Compressible
- Contract
- Density
- Diffuse
- Evaporate
- Expand
- Flow
- Gas
- Incompressible
- Liquid
- Matter
- Particles
- Particle theory
- Pressure
- Solid
- Vibrate
- Volume

Key Words Exercise

Match each key word with its meaning.

Attraction	Force that holds matter together
Compressible	The model used to explain why solids, liquids and gases behave as they do
Contract	To move up and down
Density	Get smaller
Diffuse	Can be squashed
Evaporate	Equal to mass divided by volume
Expand	Can't be squashed
Incompressible	To mix thoroughly
Particle theory	Equal to force divided by area
Pressure	A change of state from liquid to gas
Vibrate	Get larger

Comprehension

Read the passage about mercury and then answer the following questions.

1. What is the chemical symbol for mercury?
2. How did mercury get its symbol?
3. Why is mercury a very unusual metal?
4. How is mercury obtained?
5. Why is mercury used in thermometers?
6. Why is the use of mercury being phased out?
7. What might a dentist use mercury for?
8. Why were alchemists interested in mercury?
9. How did the phrase 'mad as a hatter' come about?

The element mercury is represented by the symbol Hg. The symbol comes from mercury's Latin name of *hydragyrum* (water silver). Mercury can be easily extracted from cinnabar (a mineral) by gentle heating. Mercury is a very special metal because it is a dense, silvery liquid at room temperature.

In ancient times mercury was thought to have magical properties. Alchemists believed that mercury could be transformed into other metals, including gold. Although this wasn't true, many scientists spent a lot of time trying to make it happen.

Traditionally, mercury was used in thermometers. Because mercury is a liquid, it expands and contracts more than a solid when it's heated or cooled. However, because mercury is poisonous and less dangerous alternatives have been discovered, its use in thermometers is being phased out.

Mercury was also used in hat making. People working in this industry often suffered from mercury poisoning, which can cause tremors, dementia and even death. The high numbers of hat workers suffering from these symptoms gave rise to the term 'mad as a hatter'.

Mercury will dissolve other metals like silver and gold to form amalgams. Amalgams are used in dentistry to make fillings, which are thought to be safe.

Particles

Testing Understanding

1 **Fill in the missing words to complete the sentences about matter.**

a) Everything is made up of very small pieces of matter called _____.

b) Solids have a fixed shape and a fixed _____. The particles in a solid are
_____ together. The particles can't be pushed closer together so solids
are _____. The particles in a solid can't move but they can
_____.

c) Liquids have a fixed volume but they don't have a fixed _____. The
particles in liquids are close together but they can move so liquids can _____.

d) Gases don't have a fixed shape or volume. A _____ will fill any container
into which it is placed. The _____ in gases are far apart and move quickly
in all directions. The particles can be pushed closer together; gases are _____.

2 **Read the information provided and then answer the questions.**

Density is a measure of how heavy
something is for its size.

Different substances have different densities.

The table opposite shows the densities of
different substances at room temperature.

a) Use the information in the table to draw
a bar chart to compare the densities of
the different substances.

b) i) Which two substances have the
highest densities?

ii) What do these two substances have
in common?

Substance	Density (g/cm³)
Sulfur	2.07
Diamond	3.51
Phosphorus	1.82
Iron	7.86
Copper	8.92

Skills Practice

Julie wants to know if different gases have different densities.

Julie has four rubber party balloons. She fills each balloon with a different gas:
- The first balloon contains helium.
- The second balloon contains neon.
- The third balloon contains argon.
- The fourth balloon contains krypton.

Each balloon contains the same volume of gas.

Julie then lets go of each balloon and measures how far each one moves in 5 seconds.
- The first balloon rises 1.5m.
- The second balloon falls 0.6m.
- The third balloon falls 0.8m.
- The fourth balloon falls 1.3m.

1 Copy and complete the table to show Julie's results.

Gas in the Balloon	Distance Travelled by Balloon (m)
Neon	-0.6

2 Draw a bar chart to display these results.

3 What is the independent variable in this experiment?

4 What is the dependent variable in this experiment?

5 How did Julie make this a fair test?

6 Name the piece of equipment that Julie used to measure the time in this experiment.

7 The balloon containing the densest gas falls fastest. Which gas has the highest density?

Solutions

Mixtures

Mixtures are formed when the **particles** of two or more different substances intermingle (mix together) but are not **combined** or joined together.

For example...

- air is a mixture of gas particles. About 80% of air is nitrogen and about 20% is oxygen
- seawater is a mixture of water, salts (e.g. sodium chloride) and gases (e.g. nitrogen)

- most rocks are a mixture of different **minerals** (chemical compounds). For example, granite is a mixture of quartz, feldspars and mica. The exact amount of each mineral varies from one piece of granite to another.

Mixtures don't have a fixed composition and are usually quite easy to **separate**.

Solutions

Some substances (e.g. salt) **dissolve** in liquids (e.g. water) to form mixtures called **solutions**.

- The salt is called the **solute**.
- The water is called the **solvent**.
- Salt dissolves in water so it is said to be **soluble** in water.
- The mixture that's made is called a solution.

When salt is dissolved in water the salt particles and the water particles mix together. Although the salt particles can no longer be seen, they're still there.

There's still the same number of salt particles and water particles as there were at the start of the experiment. So the total mass of the salt and the water before and after the salt has dissolved is the same.

Although water is a good solvent for many substances, some substances, e.g. nail varnish, don't dissolve in water. Nail varnish is said to be **insoluble** in water.

But, nail varnish is soluble in nail varnish remover, which contains a chemical called acetone.

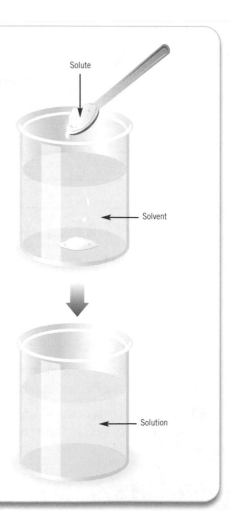

Solute

Solvent

Solution

Separating Rock Salt

Rock salt is used to grit roads in winter because it stops water from freezing to form ice.

Rock salt is a mixture of salt and sand. Salt is soluble in water but sand is insoluble.

You can use the difference in the properties of salt and sand to separate pure salt from rock salt:

1. Grind up the rock salt and place it in water. The salt will dissolve in the water to form a solution of salt water.
2. The insoluble sand will not dissolve, so you can remove it by **filtering**.
3. Place the salt water solution into an evaporating basin and leave it on a warm windowsill. The water will evaporate, leaving pure salt. The more slowly the water is allowed to **evaporate**, the larger the salt crystals will be.

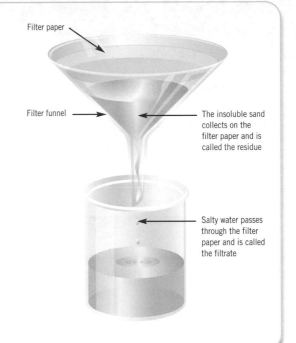

Filter paper

Filter funnel

The insoluble sand collects on the filter paper and is called the residue

Salty water passes through the filter paper and is called the filtrate

Distillation

Distillation is used to separate a liquid from a solution. You can use it to extract pure water from a solution of seawater:

- When the seawater is heated, the water evaporates to form water vapour. The salt doesn't evaporate.

- As the water vapour touches the cold surfaces of the glass condenser it **condenses** to form liquid water.
- The pure liquid is then collected in the beaker.

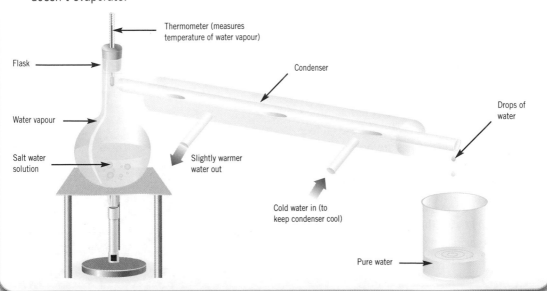

Thermometer (measures temperature of water vapour)

Flask

Condenser

Drops of water

Water vapour

Salt water solution

Slightly warmer water out

Cold water in (to keep condenser cool)

Pure water

Solutions

Chromatography

Chromatography is used to separate mixtures of different chemicals that are all soluble in a particular solvent.

For example, chromatography can be used to separate the inks found in a dye:

1. A spot of maroon dye is placed onto the filter paper.
2. The paper is then placed into a beaker containing a small amount of solvent, such as water.
3. As the solvent moves up the filter paper it carries the coloured inks with it.
4. Each ink has a slightly different solubility so it travels up the filter paper at a slightly different rate. This causes the colours to separate.
5. The chromatogram shows that the maroon dye is in fact a mixture of two different coloured inks: pink and blue.

Different dyes produce different chromatograms. The line where the ink is placed is drawn in pencil because the pencil mark is insoluble in water so it will not run and interfere with the inks being tested.

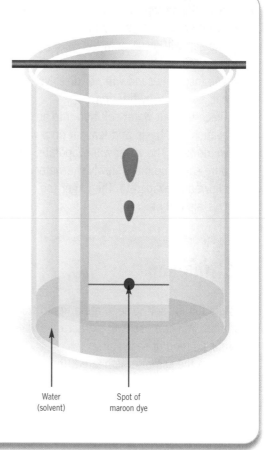

Water (solvent)

Spot of maroon dye

Using Chromatograms

Scientists use chromatography to identify substances. For example, it can be used to...

- find out if illegal drugs are present in urine samples
- identify the inks in an unknown dye.

By comparing the chromatogram from an unknown dye with the chromatogram from known inks you can see that the unknown dye contains blue and yellow inks.

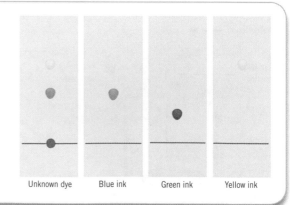

Unknown dye   Blue ink   Green ink   Yellow ink

Saturated Solutions

Solutes that dissolve very well are described as having a high solubility.

You can compare the solubility of different solutes by measuring how many grams of the solid dissolve in 100g of the solvent at a given temperature.

If you keep adding solute to a solution, eventually no more solute will dissolve and a **saturated solution** will be formed.

Factors Affecting Solubility

The amount of solute that dissolves in a solvent depends on…
- the solute used, for example, more salt than sugar will dissolve in water
- the solvent used, for example, more salt will dissolve in water than in alcohol
- the temperature used, for example, more salt will dissolve in water at a higher temperature.

If a saturated solution is warmed up it will no longer be saturated and more solute may be added.

If a saturated solution is cooled down the solution can no longer dissolve as much solute, so crystals will start to appear.

How quickly a solute dissolves can be increased by…
- increasing the temperature – as the temperature increases, the particles move faster so the solute dissolves more quickly
- stirring – if the mixture is stirred it helps the particles to mix together so the solute dissolves faster
- reducing the size of pieces – smaller pieces of solute have a larger surface area in relation to their volume, which means the particles of the solute and the solvent will mix together more quickly, making the solute dissolve faster.

Quick Test

1 Is this statement true or false? Mixtures are hard to separate.
2 What word is used to describe a substance that will not dissolve in a solvent?
3 What method can you use to obtain pure water from sea water?
4 Give an example of what chromatography can be used for.
5 Give one way in which you can make a solute dissolve more quickly.

KEY WORDS
Make sure you understand these words before moving on!
- Chromatography
- Combined
- Condense
- Dissolve
- Distillation
- Evaporate
- Filtering
- Insoluble
- Mineral
- Mixture
- Particle
- Rock salt
- Saturated solution
- Separate
- Soluble
- Solute
- Solution
- Solvent

Solutions

Match each key word with its meaning.

Chromatography	Used to separate an insoluble solid from a solution
Condense	To mix a substance with a solvent to make a solution
Dissolve	A solid that dissolves to make a solution
Distillation	A technique used to separate different coloured inks
Evaporate	A process used to separate a solvent from a solution
Filtering	A term used to describe substances that dissolve
Saturated solution	A change of state from gas to liquid
Soluble	A change of state from liquid to gas
Solute	Formed when a solute dissolves in a solvent
Solution	A solution that can't hold any more solute at a particular temperature
Solvent	A liquid that dissolves a solute to make a solution

Comprehension

Read the passage about dry cleaning and then answer the following questions.

1 Why should some clothes be dry cleaned?

2 Why is this method known as 'dry cleaning'?

3 a) Name the solvent used in dry cleaning.

b) From this passage how do you know that this solvent is a good solvent for oils?

c) Give a disadvantage of using this solvent.

Some clothes would be damaged if they were to be washed in water. Instead, they must be dry cleaned. It is called dry cleaning because no water is involved. This means that the clothes don't shrink, change colour or lose their shape.

During dry cleaning, the item is placed into a large machine with a solvent called perchloroethylene. This solvent is very good at dissolving greasy and oily stains. However, perchloroethylene is a toxic chemical and there are environmental and health concerns about its use. Even the small traces of the chemical left on clothes can irritate some people's eyes and throat.

The clothes are moved around by the machine in a similar way to a washing machine. The solvent is then removed so that it can be reused.

Testing Understanding

1 Fill in the missing words to complete the sentences about mixtures.

a) Mixtures are formed when the _____ of two or more different substances are mixed together.

b) Mixtures do not have a _____ composition and are usually quite easy to _____. Air is a mixture of different _____ including nitrogen and oxygen. _____ is a mixture of different minerals including quartz, feldspars and micas.

c) When a substance like sugar dissolves in a liquid like water a special kind of mixture called a _____ is made. The substance that dissolves is called the _____. The liquid that dissolves the substance is called the _____.

d) Water is a good _____ for many substances, but some substances like biro ink do not dissolve in water. This ink is _____ in water. But, it does dissolve in other _____ like alcohol.

2 Read the information then answer the questions that follow.

Copper sulfate is a type of salt. The table below shows the mass of copper sulfate that dissolves in 100cm³ water at different temperatures.

Temperature (°C)	Solubility (g per 100cm³ water)
0	15
20	20
40	30
60	40
80	60
100	80

a) Draw a graph to display the information shown in the table.

b) How does temperature affect the solubility of copper sulfate?

Solutions

Skills Practice

Ed wanted to see if changing the temperature of water affected how quickly salt dissolved in.

In each experiment he used 100cm³ water and 1g salt.

Here are his results:
- At 20°C it took 18 seconds.
- It took 13 seconds at 35°C.
- At 50°C it took 8 seconds.

1 Copy this table and use Ed's results to complete it.

Temperature (°C)	Time for Salt to Dissolve (s)

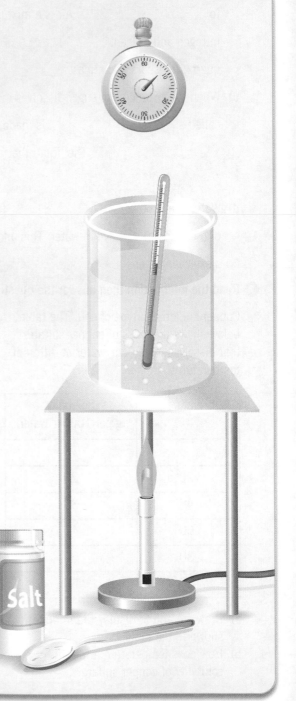

2 Draw a line graph to display Ed's results.

3 What is the independent variable?

4 What is the dependent variable?

5 Name one thing that Ed controlled in his experiment to make it a fair test.

6 What piece of equipment did Ed use to measure the volume of water?

7 Write a conclusion for Ed's experiment.

Atoms and Elements

Materials

Everything around us is made from different materials. This car is made from lots of different materials.

Different materials have different properties. The material you choose for a particular job depends on many factors including the properties you need, the cost and availability.

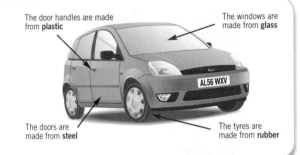

The door handles are made from **plastic**

The windows are made from **glass**

AL56 WXV

The doors are made from **steel**

The tyres are made from **rubber**

Elements

Materials are made from very small particles called **atoms**. Some materials are special because they only contain one type of atom. These materials are called **elements**.

Water pipes are often made from copper. Copper is an element because it only contains copper atoms. The atoms have a regular arrangement so it's a solid.

There are only about 100 different elements. They're displayed in the **periodic table**.

Saucepans are often made from steel. Steel isn't an element. It's a mixture of iron, carbon and chromium. Mixtures of metals are called **alloys**.

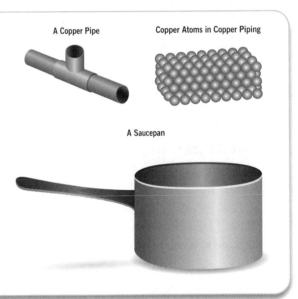

A Copper Pipe

Copper Atoms in Copper Piping

A Saucepan

Inside Atoms

Atoms are made up of three types of particles:
- Protons.
- Neutrons.
- Electrons.

Protons and neutrons are found in the nucleus of the atom. The number of protons in an atom is called the **atomic number**.

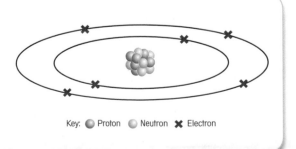

Key: ● Proton ● Neutron ✗ Electron

Atoms and Elements

The Periodic Table

Russian scientist Dimitri Mendeleev designed the periodic table, which places the elements into a meaningful order. The elements are arranged in order of increasing atomic number.

This means that elements with similar properties are in the same column or 'group'. The table also splits the elements into metals and non-metals.

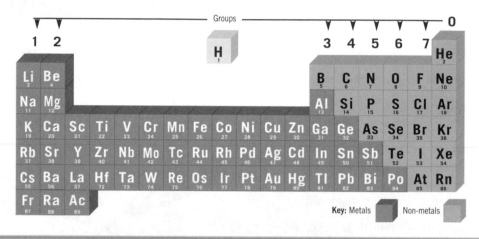

Metals and Non-Metals

The metal elements are found on the left-hand side of the periodic table. More than three-quarters of the elements are metals and they include aluminium, iron and copper.

Metals have these characteristic properties:
- They're shiny, especially when freshly cut.
- They're hard and strong.
- They're dense (heavy for their size).
- They're good conductors of heat and electricity.
- They have high melting points so they're solid at room temperature (except mercury, which is a liquid).
- They can be drawn into wire and hammered into shape.
- A few metals (e.g. iron, cobalt and nickel) are magnetic but most are non-magnetic.

The non-metal elements are found on the right-hand side of the periodic table. Less than a quarter of the elements are non-metals and they include helium, oxygen and carbon.

Non-metals have these characteristic properties:
- They have a low density (are light for their size).
- They're poor conductors of heat and electricity (except carbon in the form of graphite, which is a good electrical conductor).
- They have lower melting points so many are gases (e.g. nitrogen) at room temperature. A few are solid (e.g. sulfur). Only bromine is a liquid.

Symbols

Scientists use **symbols** to represent different elements. In some cases, the symbol is simply the first letter of the element's name.

In other cases, where several elements start with the same letter, the symbol is the first letter of the element's name followed by another letter from its name. When two letters are used...

- the first letter is a capital letter
- the second letter is lower case.

Sometimes the symbol comes from the Latin name.

Element	Symbol
Carbon	C
Nitrogen	N
Oxygen	O
Calcium	Ca
Chlorine	Cl

Element	Latin Name	Symbol
Iron	ferrum	Fe
Sodium	natrium	Na

Compounds

A **compound**...

- is a substance formed when the atoms of two or more different elements are joined together by a chemical reaction
- has very different properties from the elements it was made from.

Molecules

A **molecule** is formed when a small group of atoms are joined together.

The atoms can be the same (e.g. hydrogen) or different (e.g. carbon dioxide).

Dalton's Symbols

The scientist John Dalton used different symbols to represent atoms.

 Oxygen   Carbon   Hydrogen   Sulfur

This diagram represents an element (carbon). The atoms are close together and are in a regular arrangement, so this is solid carbon. 	This diagram represents a compound (sulfur dioxide). Each sulfur atom is joined to two oxygen atoms. The molecules are far apart, so this is sulfur dioxide gas.
This diagram shows molecules of an element (oxygen). The molecules are far apart, so this is oxygen gas. 	This diagram also represents a compound (water). Each oxygen atom is joined to two hydrogen atoms. The molecules are far apart, so this is water vapour.

Atoms and Elements

Reactants and Products

During a chemical reaction, new substances are made. The chemicals present at the start of the reaction are called the **reactants**.

The new substances made by the reaction are called the **products**. Here are two examples:

1 Hydrogen and Oxygen	**2** Carbon and Oxygen
When hydrogen is burned, it reacts with oxygen to form water vapour. Hydrogen and oxygen are the reactants; water is the product.	When carbon is burned in a good supply of oxygen, a chemical reaction takes place and carbon dioxide is made.

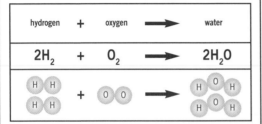

hydrogen + oxygen → water	carbon + oxygen → carbon dioxide
$2H_2 + O_2 \rightarrow 2H_2O$	$C + O_2 \rightarrow CO_2$

Metal Oxides

Most metals react if they're heated in air. The metal reacts with the oxygen in air to form a metal **oxide**. Metals burn more vigorously in pure oxygen than they do in air. This is the reaction between magnesium and oxygen:

Magnesium

Combustion spoon

Bunsen burner

The heated magnesium burns brightly with a white flame in a gas jar full of oxygen.

Magnesium oxide is a whitish powder.

Quick Test

1 What is special about an element?
2 How are elements arranged in the periodic table?
3 An element is solid, conducts electricity and is shiny when freshly cut. What does this tell you about the element?
4 Identify the element from the symbol...
   a) C b) O c) Ca d) Fe
5 Name the product made when magnesium metal is burned in air.

Key Words Exercise

Match each key word with its meaning.

Alloy	A material that's made of only one type of atom
Atom	A very small particle
Atomic number	A way of displaying elements in order of increasing atomic number
Compound	A mixture of metals
Element	A one or two-letter code used to represent an element
Molecule	Contains atoms of two or more different elements that have been joined together by a chemical reaction
Oxide	A small group of atoms that are joined together
Periodic table	The chemical used up during a reaction
Product	The number of protons in an atom
Reactant	The substance made by a chemical reaction
Symbol	Made when a material is burned in air

Comprehension

Read the passage about Dimitri Mendeleev and the periodic table, then answer the following questions.

1. Where was Dimitri Mendeleev born?

2. In which city did Mendeleev study science?

3. How did Mendeleev become famous?

4. How many elements were known when Mendeleev designed his table?

5. Why did he leave gaps in his table?

6. Why didn't Mendeleev include any noble gases in his table?

7. Which element was named after Mendeleev?

8. What else was named after Mendeleev?

Dimitri Mendeleev was born in Siberia in 1834. He was the youngest of at least 14 children. After his father died and his mother's glass factory burned down, the family moved to St. Petersburg. Mendeleev studied science at the university and eventually became a professor.

Mendeleev became famous for designing the first version of the periodic table. He placed the 63 known elements in order of increasing atomic weight. He left spaces and used his table to make detailed predictions about the properties of the missing elements. When the elements were eventually discovered and their properties closely matched Mendeleev's predictions, he proved that the table was a powerful way of helping us to understand the world around us.

Mendeleev also changed the order of the elements occasionally so that elements with similar properties were placed in the same vertical column or group. This meant that Mendeleev had actually placed the elements in order of increasing atomic number. Mendeleev's table looked a little different from the modern periodic table. For example, he didn't include any noble gases as they hadn't been discovered and many scientists have made improvements to his original ideas.

Today, more than 100 years after his death, Mendeleev is famous all over the world. The radioactive element 101 was named mendelevium and the Mendeleev crater on the Moon was named after him.

Atoms and Elements

Testing Understanding

1 **Fill in the missing words to complete the sentences about atoms and elements.**

a) Everything around us is made of small particles called _____. Some materials are made of only one type of atom and these materials are called _____.

b) There are about 100 different elements and they're often arranged in the _____ table. The elements can be split into metals and _____-metals. The metals are found on the _____-hand side of the periodic table.

c) Metals are hard and _____, especially when they're freshly cut. Only one metal isn't solid at room temperature. This metal is called _____ and is used in thermometers.

d) Non-metals are poor conductors of both heat and _____. Oxygen is a non-metal element. When magnesium is heated in air, it reacts with _____ to produce magnesium _____.

2 **Study the diagrams, then answer the questions that follow.**

These symbols represent atoms:

These diagrams show different combinations of these atoms:

A	B	C	D

a) Give the letters of the diagrams that show elements.
b) Give the letter of the diagram that shows a solid.
c) Give the letter of the diagram of an element that's a gas.
d) Give the letters of the diagrams that show compounds.

Skills Practice

Duncan wants to investigate what happened to the mass of magnesium metal when it was burned in oxygen.

1 Complete the word equation to sum up the reaction between magnesium and oxygen.

magnesium + oxygen ➡ ....

2 Duncan measured the mass of the magnesium at the start of the reaction and the mass of the product at the end of the reaction. The results are shown in the table below.

  a) Complete the table by adding the units at the top of the last column.

  b) Calculate the difference in mass and write this in the last column of the table.

3 a) Use the information in the table to complete the graph below.

  b) Add a line of best fit.

  c) Use the graph to write a conclusion for the experiment.

Mass of Magnesium Burned (g)	Mass of the Magnesium After it's Burned (g)	Difference in Mass
0.6	1.0	0.4
1.2	2.0	
1.8	3.5	
2.4	4.0	

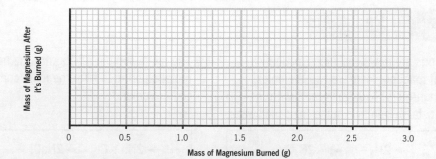

Compounds and Mixtures

How Atoms Combine to Form Compounds

You will probably remember that...
- an **element** is made of only one type of **atom**
- **compounds** have atoms of two or more different elements joined together chemically.

There are two basic types of compound:
- An ionic compound is created when a metal reacts with a non-metal. Magnesium oxide is an ionic compound made when magnesium atoms combine with oxygen atoms.
- A covalent compound is formed when two or more non-metals react together. Carbon dioxide is a covalent compound. Each carbon dioxide molecule consists of one carbon atom combined with two oxygen atoms.

The Structure of Magnesium Oxide

The particles have a regular arrangement, so it's a solid.

The Structure of Carbon Dioxide

Carbon dioxide consists of a small group of atoms, so it's a molecule. The molecules are far apart, so this represents a gas.

Formulae

Scientists use formulae to represent compounds. A chemical **formula** shows the type and ratio of atoms present.

Name	Formula	Composition
Bromine	Br_2	Two atoms of bromine per molecule
Sulfur dioxide	SO_2	One atom of sulfur and two atoms of oxygen per molecule
Copper carbonate	$CuCO_3$	One copper atom to one carbon atom to three oxygen atoms

Chemical Reactions

Compounds are new substances made by a **chemical reaction**. Here are two examples:

❶ The non-metal hydrogen reacts with the non-metal oxygen to form the compound water.

Two molecules of hydrogen... → $2H_2 + O_2$ → $2H_2O$ ...to form two molecules of water
...react with one molecule of oxygen...

❷ The metal sodium reacts with the non-metal chlorine to form the ionic compound sodium chloride.

Two sodium atoms... → $2Na + Cl_2$ → $2NaCl$ ...to form sodium chloride that has sodium and chlorine in the ratio 1:1.
...react with one chlorine molecule...

The Signs of a Chemical Reaction

Compounds can react chemically and when they do there are often 'tell-tale' signs that a chemical reaction has taken place.

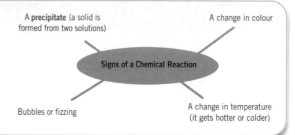

A **precipitate** (a solid is formed from two solutions)

A change in colour

Signs of a Chemical Reaction

Bubbles or fizzing

A change in temperature (it gets hotter or colder)

Chemical Reactions Involving Compounds

Here are some examples of chemical reactions involving compounds:

1 Sucrose is a type of sugar. When it's heated, a chemical reaction takes place:

sucrose $\longrightarrow$ water vapour + carbon

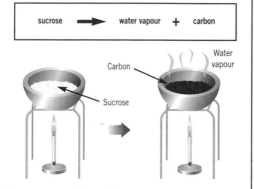

Carbon

Water vapour

Sucrose

2 When pieces of magnesium carbonate are added to a solution of hydrochloric acid, a chemical reaction takes place:

magnesium carbonate + hydrochloric acid $\longrightarrow$ magnesium chloride + water + carbon dioxide

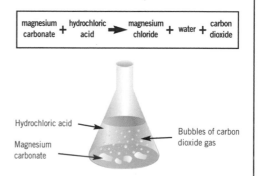

Hydrochloric acid

Magnesium carbonate

Bubbles of carbon dioxide gas

3 When a solution of iron chloride is added to a solution of sodium carbonate, a chemical reaction takes place:

sodium carbonate + iron chloride $\longrightarrow$ sodium chloride + iron carbonate (precipitate)

The iron carbonate is formed as a precipitate, an insoluble solid made when two solutions react together.

Iron chloride solution

Sodium chloride solution

Sodium carbonate solution

Iron carbonate precipitate

4 When a solution of ammonium hydroxide is added to a solution of copper sulfate, a chemical reaction takes place:

ammonium hydroxide + copper sulfate $\longrightarrow$ ammonium sulfate + copper hydroxide (precipitate)

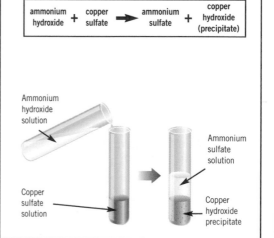

Ammonium hydroxide solution

Copper sulfate solution

Ammonium sulfate solution

Copper hydroxide precipitate

Compounds and Mixtures

Mixtures

In a **mixture**, two or more different substances are mixed together, but not chemically joined.

Mixtures...
- don't have a fixed composition (i.e. they don't need to contain a fixed number of atoms and molecules)
- are easy to separate.

Compounds...
- do have a fixed composition
- are much harder to separate.

A mixture of iron and sulfur can be changed by adding more iron or more sulfur. The mixture can be separated by using a magnet. The magnet attracts the iron, which is **magnetic**.

If the mixture of iron and sulfur is heated gently, the iron atoms and the sulfur atoms become chemically joined. A new compound called iron sulfide (FeS) is formed. It has one iron atom for every sulfur atom. Iron sulfide has different properties from iron and sulfur. It's a black, non-magnetic powder.

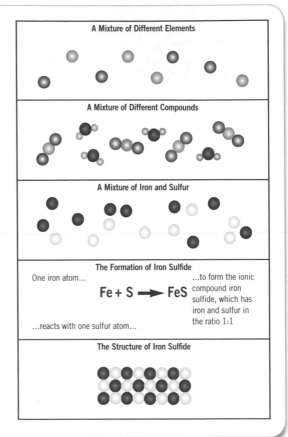

A Mixture of Different Elements

A Mixture of Different Compounds

A Mixture of Iron and Sulfur

The Formation of Iron Sulfide

One iron atom...

$$Fe + S \longrightarrow FeS$$

...reacts with one sulfur atom...

...to form the ionic compound iron sulfide, which has iron and sulfur in the ratio 1:1

The Structure of Iron Sulfide

Everyday Mixtures

Here are some examples of everyday mixtures:
- Seawater, which is a mixture of water, dissolved salts and gases.
- Rocks, most of which are a mixture of different minerals. For example, granite is a mixture of feldspar, quartz and mica minerals.
- Mineral water, which is a mixture of water and dissolved salts. The amount and type of dissolved salts varies depending on the type of rocks the water has flowed through.
- Air, which is a mixture of different gases.

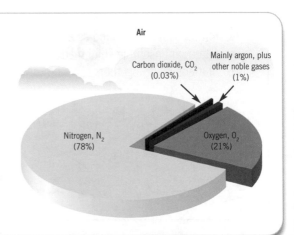

Air

Carbon dioxide, CO_2 (0.03%)

Mainly argon, plus other noble gases (1%)

Nitrogen, N_2 (78%)

Oxygen, O_2 (21%)

Fractional Distillation

Air can be separated by **fractional distillation**:
- First the air is cooled to -200°C. At this temperature, all the gases in air **condense** to form liquids.
- The liquid air is then placed in a fractional distillation column and is warmed up.
- As each liquid boils at a different temperature, the various gases can be collected and removed.

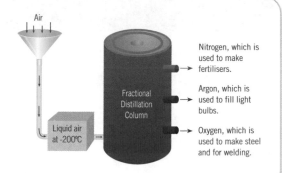

Melting and Boiling Points

When substances are heated or cooled they can change state, such as from solid to liquid or liquid to gas.

Pure elements and compounds melt and boil at fixed temperatures. Iron has a melting point of 1535°C and a boiling point of 2750°C. This means that...
- below 1535°C iron is a solid.
- between 1535°C and 2750°C iron is a liquid
- above 2750°C iron is a gas.

Because pure elements and compounds melt and boil at fixed temperatures, you can use melting points and boiling points to identify unknown substances.

Mixtures don't have fixed melting points and boiling points. Pure water boils at 100°C. If salt is added to pure water, a mixture is formed. The boiling point of the salty water is higher than pure water. The more salt that's added, the higher the boiling point becomes.

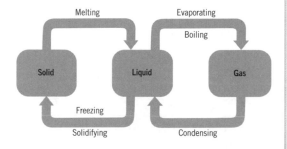

Quick Test

1. Name the compound formed when magnesium reacts with oxygen.
2. Name the compound formed when sodium reacts with chlorine.
3. A molecule of carbon dioxide has the formula CO_2. Explain what this formula means.
4. How can air be separated?

Compounds and Mixtures

Key Words Exercise

Match each key word with its meaning.

Atom	A change in which new substances are made
Chemical reaction	A substance made of only one type of atom
Compound	A very small particle
Condense	Contains atoms of two or more different elements that are chemically joined
Element	A word describing a material that's attracted by a magnet
Formula	A way to separate mixtures of liquids that have different boiling points
Fractional distillation	Two or more different substances that are mixed together but not chemically joined
Magnetic	Turn from a gas to a liquid
Mixture	An insoluble solid made when two solutions react together
Precipitate	A code to represent the type and number of atoms present

Comprehension

Read the passage about nitrogen, then answer the following questions.

1. How is nitrogen obtained?

2. Describe an ammonia molecule.

3. What is a catalyst?

4. Why are fertilisers used?

5. Why is nitrogen needed by plants?

6. How can you tell that a plant doesn't have enough nitrogen?

Nitrogen is obtained by the fractional distillation of liquid air. Nitrogen is heated with hydrogen to produce the compound ammonia, NH_3. Each ammonia molecule contains one nitrogen atom and three hydrogen atoms. An iron catalyst is also used. A catalyst is a special chemical that increases the rate of reaction but isn't used up during the reaction. This means that a catalyst can be reused many times.

Ammonia is used to produce nitrogen fertilisers. Gardeners and farmers use fertilisers to help their plants grow better. Fertilisers help plants by replacing the nutrients that the plants use up as they grow. Nitrogen fertilisers are very important. Nitrogen makes plants grow better and produce lots of lush green leaves. If there isn't enough nitrogen, plants become stunted and start to turn yellow. This means that these plants will not produce high yields of crops.

Testing Understanding

1 **Fill in the missing words to complete the sentences about compounds and mixtures.**

   a) _____, like copper, are made of only one type of atom. Compounds are

     made when _____ of two or more different elements are joined together

     by a chemical _____ .

   b) The compound magnesium oxide is made when magnesium reacts with _____ .

     The compound carbon dioxide is made when _____ reacts with oxygen.

   c) Compounds have a fixed _____ and are _____ to separate.

   d) A _____ is formed when two or more substances are mixed together.

     Mineral water and most rocks are examples of everyday _____ .

   e) Mixtures don't have a fixed _____ and are _____

     to separate.

   f) Air is a mixture. The main gases in air are _____ and oxygen. Air can be

     separated by the fractional _____ of liquid air.

2 **Study the diagrams, then answer the questions that follow.**

The symbols represent different types of atoms:

These diagrams show the arrangement of the atoms in four different substances:

A	B	C	D

   a) Give the letter of the diagram that shows an element.
   b) Give the letter of the diagram that shows one compound.
   c) Give the letter of the diagram that shows a mixture of elements.
   d) Give the letter of the diagram that shows a mixture of compounds.

Compounds and Mixtures

Skills Practice

Andy wants to find out if the amount of salt added to water affects its boiling point.

1 Pure water boils at 100°C. Draw an arrow on the temperature scale below to show the temperature at which pure water freezes.

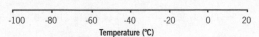

-100   -80   -60   -40   -20   0   20
Temperature (°C)

2 a) What piece of equipment should Andy use to measure the boiling point of the water?
   b) What piece of apparatus should he use to measure the volume of water used in the experiment?
   c) What piece of apparatus should he use to measure the amount of salt used in the experiment?

3 Complete the table below to show the different factors in Andy's experiment. Put one tick in each row.

	Factor to Change	Factor to Keep Same	Factor to be Measured
Temp. the Water Boils at			
Volume of Water			
Mass of Salt			

4 The table below shows the results from Andy's experiment.

Mass of Salt Added (g)	Boiling Temperature (°C)
0	100
0.5	102
1.0	102
1.5	103
2.0	104

   a) Use this information to complete the graph.
   b) Add a line of best fit.

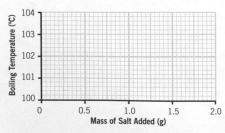

5 Andy decided to repeat the experiment. Suggest why this was a good idea.

     For further practice, go to p.20-23 of the Year 8 Science Essentials Workbook

Rocks and Weathering

What Rocks are Made of

Most rocks are mixtures of different **minerals**, naturally-occurring solid compounds with a crystalline structure. There are three types of rock:

Rock Type	Properties	Examples	
Sedimentary	These rocks are made of grains that are stuck together. They're formed in layers when **sediment** is **deposited**. They're quite soft and crumbly.	Sandstone	Limestone
Metamorphic	These rocks are made of **interlocking** crystals. They're formed when existing rocks are changed by high temperatures and pressures. They're hard, smooth and shiny, and may have layers of crystals.	Slate	Marble
Igneous	These rocks are made of interlocking crystals. They're formed when molten (liquefied) rock cools down and solidifies. The size of the crystals depends on the rate of cooling. These rocks are very hard.	Basalt	Granite

Porous and Non-Porous Rocks

In sedimentary rocks, the shape of the grains means that they can't interlock and there are gaps between them. These rocks are **porous** – there are gaps for air or water to get into.

In metamorphic and igneous rocks, the mineral crystals interlock with no holes or gaps between them. These rocks are non-porous – there are no gaps for air or water to get into.

If a porous rock is placed in water:
- Bubbles appear as air is lost from the rock.
- The mass of the rock increases as the gaps between the grains become filled with water.

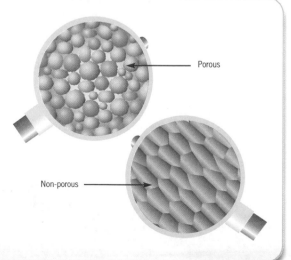

Porous

Non-porous

Rocks and Weathering

Physical Weathering

Temperature changes can break down rocks in two ways:

Type of Weathering	Description
Freeze-thaw Weathering Water	Rocks are broken down by the large forces exerted when water **freezes** to form ice. First, water gets into existing cracks in the rock. When the temperature falls to 0°C or below, the water freezes to form ice. When water freezes, it expands (gets bigger). The large forces exerted by the ice can force open the cracks in the rock, making them even bigger. Over long periods of time, this process can even break down mountains into fragments of rock.
Onion-skin Weathering	Large changes between the day-time and night-time temperatures can break down rocks. In the day, the rocks are heated and expand. During the night, the rocks cool down and contract (get smaller). Eventually, this can cause the outer layers of the rock to peel off like the skin of an onion.

Landscapes and Weathering

Scree slopes are often found at the bottom of rock cliffs. As the rock in the cliff face is weathered, fragments of rock break away. These fragments fall down and form a smooth scree slope.

Scree

How Rainwater Causes Rocks to Weather

Air contains small amounts of carbon dioxide. Carbon dioxide dissolves in rainwater to form carbonic acid, so rainwater is naturally slightly acidic. This acidic rain can cause the chemical weathering of rocks.

In polluted areas, gases, including sulfur dioxide from industrial processes and nitrogen oxides from car exhausts, may also dissolve in rainwater.

Sulfur dioxide and nitrogen oxides lower the pH of the rainwater even more to form acid rain. This speeds up the chemical weathering.

If the rainwater falls on rocks that contain metal carbonates, such as limestone, the rocks are weathered more quickly. When rocks are weathered, they become discoloured. Weathering can damage the detail on statues.

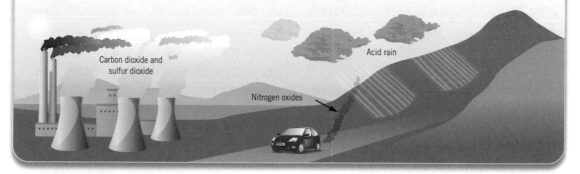

Carbon dioxide and sulfur dioxide

Acid rain

Nitrogen oxides

Transportation

Pieces of rock can be transported by **glaciers**, wind or rivers:

- Glaciers are slow-moving rivers of ice that can carry pieces of rock. When the glaciers melt, the fragments are deposited.
- When rocks are weathered, only the hardest materials (e.g. sand grains) remain. The wind can blow sand grains great distances. As the sand grains travel, they rub against each other and become more rounded.
- Rivers carry smaller fragments of rock further than larger pieces. As the pieces move, they rub against each other and against the river bed, making them smaller and more rounded. When the river can no longer carry the particles, they're deposited as layers of sediment. Grains of a similar size are deposited at the same time.

Where a River Deposits Rock Fragments

Large grains deposited here

Medium grains deposited here

Small grains deposited here

Rock fragments become more rounded as they move down the river bed.

Evaporates

Layers of sediment can also be formed when water that contains dissolved salts evaporates.

The salts are left behind as layers of solid sediment. These minerals are called **evaporates**.

Rocks and Weathering

Fossils and Evidence About the Past

Some sedimentary rocks contain **fossils**. Fossils are the remains of plants and animals that lived long ago and have been preserved in rock. Normally, the shells and bones are preserved best.

Occasionally, deformed fossils are found in metamorphic rocks but they're never found in igneous rocks. Igneous rocks formed from rock that was so hot that any remains would be destroyed.

Fossils form when dead plants and animals are covered in sediment before they can rot away. You can use fossils to help date rocks. If the same type of fossil is found in two different rocks, the rocks were formed at the same time.

Fossils can also tell you about the conditions that existed when the rocks formed. For example, if a rock contains fossilised sea shells, then it must have been formed in a marine environment.

Rock cliffs can also tell you about how rocks formed. The oldest rocks are usually found at the bottom of the cliff because they formed first. The newer rocks are found on top.

Younger rock · Fossil · Older rock

Fossil Fuels

Coal, oil and natural gas are **fossil fuels**. Coal is formed from the remains of dead plants, and oil and natural gas are formed from the remains of dead sea creatures and plants.

Fossil fuels are non-renewable – they take millions of years to form, which is much slower than the rate they're being used up.

Fossil fuels contain carbon. When they're burned, carbon dioxide is formed and this gas contributes to global warming:

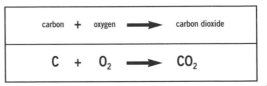

$$carbon + oxygen \longrightarrow carbon\ dioxide$$

$$C + O_2 \longrightarrow CO_2$$

Quick Test

1. Which types of rock contain interlocking crystals?
2. What happens when a porous rock is placed in water?
3. How can ice weather rock?
4. How is acid rain formed?
5. What is a glacier?
6. What happens to rocks as they're carried by a river?

KEY WORDS
Make sure you understand these words before moving on!

- Deposit
- Evaporates
- Fossil fuels
- Fossils
- Freeze
- Glacier
- Interlocking
- Mineral
- Porous
- Sediment

Key Words Exercise

Match each key word with its meaning.

Deposit	• •	The remains of dead plants and animals
Evaporates	• •	Solids that are deposited by a river
Fossil fuels	• •	Lay down
Fossils	• •	A slow-moving river of ice
Freeze	• •	Turn from liquid to solid
Glacier	• •	Joined together with no gaps
Interlocking	• •	A naturally-occurring solid compound with a crystalline structure
Mineral	• •	Minerals formed when water that contains dissolved salts evaporates
Porous	• •	Energy resources including coal, oil and natural gas
Sediment	• •	Contains gaps between grains that air or water can enter

Comprehension

Read the passage about the great fossil finder, Mary Anning, then answer the following questions.

1. Where was Mary Anning born?

2. Why did Mary and her brother start to collect fossils?

3. Why was fossil collecting a dangerous thing to do?

4. Describe an ichthyosaur.

5. Name Mary's other outstanding fossil finds.

6. How did Mary's findings show that species could become extinct?

Mary Anning was born in 1799 in the town of Lyme Regis, Dorset, on the south coast of England. From an early age Mary seemed marked out for a remarkable life. When she was a baby, she was one of four people struck by a lightning bolt. Mary was the only one to survive.

When Mary's father died of tuberculosis, Mary and her brother Joseph collected fossils from the cliffs near Lyme Regis, which they sold to earn money for the family. This was very dangerous because the cliffs were unstable. High tides and storms dislodged pieces of rock and exposed fossils, but the cliffs could easily collapse and bury the children.

A few months after her father's death, 12-year-old Mary discovered the first complete fossilised skeleton of a crocodile-like dinosaur called an ichthyosaur. She later discovered the first examples of a plesiosaur and a pterosaur.

She became known as the 'greatest fossil finder the world has ever known', even though society and science were dominated by men at the time.

Mary's discoveries were important because they provided evidence of animals that had existed in the past but weren't alive today, so they showed that species could become extinct. She gave an insight to our understanding of who we are and our place in the universe. The tongue twister 'she sells sea-shells on the sea shore' keeps her memory alive.

Rocks and Weathering

Testing Understanding

1 **Fill in the missing words to complete the sentences about rocks and weathering.**

   a) Most _____ are mixtures of different minerals. Sandstone and limestone are _____ rocks. They're made of grains that are stuck _____.

   b) Marble and slate are _____ rocks. Granite and basalt are _____ rocks. Both of these types of rock have interlocking _____. There are no _____ between the crystals so these rocks aren't porous.

   c) Water can weather rocks. Water gets into existing _____ and when it freezes it changes state from liquid to _____. As the ice forms, it _____ and pressure builds up on the cracks, causing them to be pushed _____ apart. Temperature changes can also cause _____-skin weathering.

   d) Normal rainwater is slightly acidic because it contains trace amounts of the gas, carbon _____. This _____ the pH of the rainwater. Rainwater weathers all rocks but rocks containing metal _____ weather faster.

   e) In polluted areas, sulfur dioxide can dissolve in rainwater to form _____ rain. This reduces the _____ of the acid even more and speeds up the rate of chemical weathering.

2 **Study the diagram opposite of a cliff face, then answer the questions.**

   a) Which is the oldest type of rock? Explain your answer.

   b) The limestone contains a fossil of a seashell. What is a fossil?

   c) What does this fossil tell you about the conditions in which this limestone was formed?

   d) An identical fossil was found in another type of rock further down the coast. What does this tell you about these two rocks?

   e) Explain how ice can cause the weathering of the rocks in this cliff face.

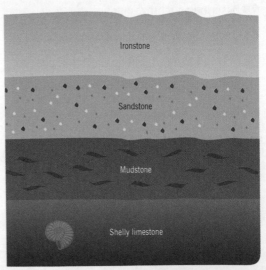

Ironstone

Sandstone

Mudstone

Shelly limestone

Skills Practice

Ed uses a pH meter to measure the pH of four water samples.

The table below shows his results.

Water Sample	pH of the Solution
A	7.0
B	5.2
C	6.4
D	4.8

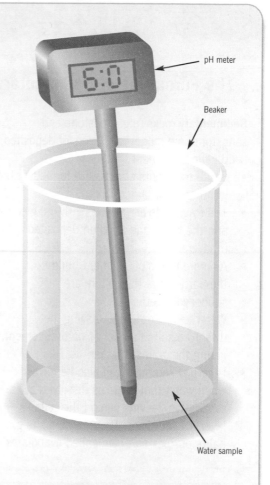

pH meter

Beaker

Water sample

1. Complete the bar chart below to show the pH of each water sample.

2. Suggest an advantage of using a pH meter rather than universal indicator paper to measure the pH of the water samples.

3. Water sample C is slightly acidic because carbon dioxide has dissolved in the rainwater. Carbon dioxide is formed when carbon is burned in oxygen. Complete the word equation below to sum up this reaction.

carbon + ... ➡ ....

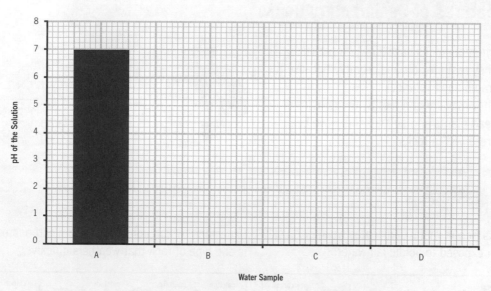

pH of the Solution

Water Sample

The Rock Cycle

The Formation of Sedimentary Rocks

Sedimentary rocks (e.g. limestone and sandstone) are made from layers of **deposited** solid called sediment:

- Small fragments of rocks are formed by the weathering of larger rocks.
- The fragments are moved by rivers, glaciers and the wind, and are then deposited in layers.
- As more layers of sediment build up, the particles at the bottom are squeezed together.
- Water is squeezed out, leaving the dissolved salts that act as natural **cement**, gluing the grains together.
- Layers of sediment can also be formed when water that contains dissolved salts evaporates, leaving behind the salts as layers of sediment. These minerals (e.g. halite or 'rock salt') are called **evaporates**.

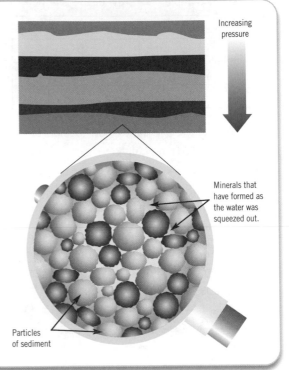

Increasing pressure

Minerals that have formed as the water was squeezed out.

Particles of sediment

Features of Sedimentary Rocks

Sedimentary rocks have these properties:

- They're quite soft and crumbly because the grains don't interlock.
- They may contain **fossils**, the remains of dead plants and animals that have been preserved in rocks.

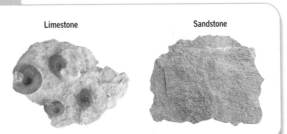

Limestone

Sandstone

Limestone

Rocks rich in metal carbonates weather quickly when exposed to acidic rainwater:

metal carbonate + acid ➡ salt + water + carbon dioxide

Limestone, which contains calcium carbonate, is one type of rock that weathers quickly:

calcium carbonate + sulfuric acid ➡ calcium sulfate + water + carbon dioxide

The Formation of Metamorphic Rocks

Metamorphic rocks (e.g. **marble** and **slate**) are made when high temperatures and pressures change existing (usually sedimentary) rocks:

- Existing rocks are heated by molten rock called magma. As the rocks are heated, they recrystallise and new crystals are formed.
- Existing rocks are subjected to high pressures by forces in the Earth's crust or by the weight of rocks above them.

Original Rock	Metamorphic Rock
Mudstone	Slate
Limestone	Marble
Sandstone	Quartzite

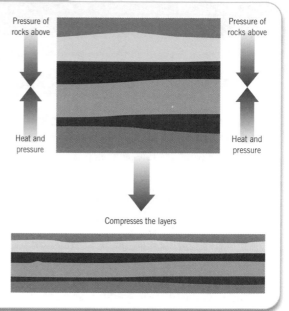

Pressure of rocks above

Heat and pressure

Pressure of rocks above

Heat and pressure

Compresses the layers

Features of Metamorphic Rocks

Metamorphic rocks have these properties:

- They're normally hard and shiny.
- They sometimes have bands of crystals.
- Different minerals form at different temperatures, so they can tell you how far the original rock was from the heat source that changed it.
- They occasionally contain fossils that have been deformed (changed) by low-level or 'low-grade' metamorphism. Higher level or 'high-grade' metamorphism can destroy the fossils completely.

Marble

Bands of Crystals in Gneiss

Folding and Faulting

When layers of rock are subjected to large forces, folding can be seen. Folding can create mountain chains, such as the Himalayas.

A fault occurs if the forces in the Earth's crust cause the layers of rock to fracture.

How Folding Affects Layers of Rock

Syncline	Anticline

The Rock Cycle

The Formation of Igneous Rocks

Igneous rocks form when molten rock cools and solidifies. The size of the crystals present in the rocks depends on the rate of cooling.

Molten rock...
- below the Earth's surface is called **magma**
- above the Earth's surface is called **lava**.

Magma in a magma chamber is slow to cool because there's a lot of it and it's well insulated by the surrounding rock. Igneous rocks formed in these conditions (e.g. granite) have big crystals because the slow rate of cooling gives plenty of time for the crystals to grow.

Magma can cut through rocks. The magma cools and solidifies into igneous rock. This shows the igneous rock is younger than the rock it cuts into.

At the Earth's surface, lava cools quickly because it's in contact with air or water. Igneous rocks formed in these conditions (e.g. basalt) have small crystals because the fast cooling gives less time for the crystals to grow.

Pumice is an igneous rock formed when volcanoes explode. Lava rich in volcanic gases is thrown out and it quickly solidifies to leave a rock that has lots of holes where the gases were.

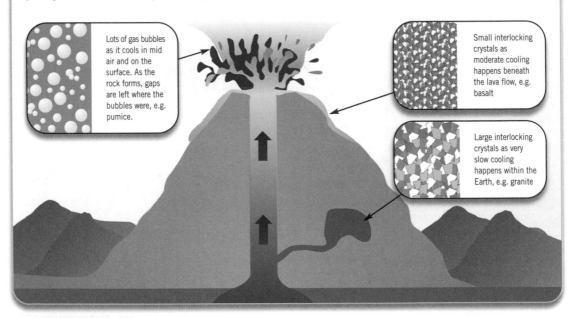

Lots of gas bubbles as it cools in mid air and on the surface. As the rock forms, gaps are left where the bubbles were, e.g. pumice.

Small interlocking crystals as moderate cooling happens beneath the lava flow, e.g. basalt

Large interlocking crystals as very slow cooling happens within the Earth, e.g. granite

Features of Igneous Rocks

Igneous rocks have these properties:
- They have interlocking crystals. This means that they're very hard and don't weather easily.
- They never contain fossils as the molten rock would destroy any organisms.

Basalt

Granite

The Rock Cycle

The rock cycle explains how the atoms in rocks are being recycled over long periods of time:

- When molten rock cools down and solidifies, igneous rocks are formed.
- Igneous rocks are weathered and the rock fragments are transported by rivers, glaciers or the wind. Eventually they're deposited.
- As the layers of sediment build up, the lowest layers are compressed and water is squeezed out. The grains become cemented together and sedimentary rocks are formed.

- Heat and pressure can change the sedimentary rocks and some igneous rocks, forming metamorphic rocks.
- If the temperature and pressure are increased further, the metamorphic rocks can melt completely to form magma.
- In this way, the atoms are being constantly recycled and an atom that is today in a sedimentary rock will, one day, be in a metamorphic rock or an igneous rock.

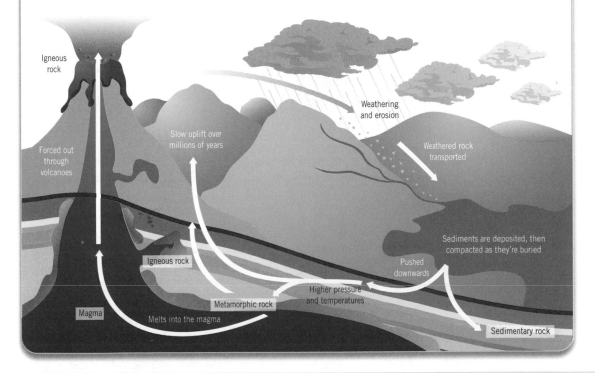

Igneous rock

Forced out through volcanoes

Slow uplift over millions of years

Weathering and erosion

Weathered rock transported

Igneous rock

Pushed downwards

Sediments are deposited, then compacted as they're buried

Metamorphic rock

Higher pressure and temperatures

Magma

Melts into the magma

Sedimentary rock

Quick Test

1. Which is the softest type of rock?
2. Name two things that can cause metamorphism?
3. Name the metamorphic rock formed from mudstone.
4. What is molten rock below the Earth's surface called?
5. What is molten rock above the Earth's surface called?
6. An igneous rock contains large crystals. Explain what this tells you about the formation of this rock.

KEY WORDS
Make sure you understand these words before moving on!
- Cement
- Deposit
- Evaporates
- Fossils
- Lava
- Magma
- Marble
- Sedimentary
- Slate

The Rock Cycle

Key Words Exercise

Match each key word with its meaning.

Key Word	Meaning
Cement	Molten rock above the Earth's surface
Deposit	Molten rock below the Earth's surface
Evaporates	Lay down
Fossils	Minerals formed when water that contains dissolved salts evaporates
Lava	The remains of dead plants and animals
Magma	Type of metamorphic rock formed from limestone
Marble	Type of metamorphic rock formed from mudstone
Sedimentary	Type of rock formed when sediments are cemented together by dissolved salts
Slate	'Glue' that sticks sediments together

Comprehension

Read the passage about the Ring of Fire, then answer the following questions.

1. Which ocean is surrounded by the Ring of Fire?

2. Which features are associated with the Ring of Fire?

3. What is the centre of the Earth called?

4. How fast do the Earth's plates move?

5. How are earthquakes formed?

6. How are volcanoes formed?

Most of the world's earthquakes and volcanoes occur in a horseshoe-shaped band around the Pacific Ocean called the 'Ring of Fire'. Scientists have discovered that the Earth has a layered structure. At the centre of the Earth is the core. This is surrounded by a middle layer called the mantle. The outside layer, on which we live, is called the crust. The crust and the upper part of the mantle are split into about a dozen pieces called plates. These plates move across the surface of the Earth at the rate of a few centimetres per year, which is about the rate at which your finger nails grow.

The Ring of Fire marks the boundaries between many of these plates. The plates can move in three ways. They can move past each other, towards each other or away from each other. When the plates move past each other, they occasionally get stuck. Gradually, the forces build up until suddenly the plates move and the energy that had built up is released as an earthquake.

When the plates move towards each other, one plate is forced underneath the other. This results in volcanoes. When the plates move away from each other, magma comes to the surface. This normally happens under the sea and new sea floor is made. These plate movements result in the large number of earthquakes and volcanoes observed around the Ring of Fire.

Testing Understanding

1 Fill in the missing words to complete the sentences about the rock cycle.

a) Sedimentary _____ are formed from layers of sediment. As more layers build up, the _____ increases and water is squeezed out. The sediments are cemented together by the dissolved _____.

b) Sedimentary rocks, like sandstone and limestone, are quite soft and _____.

c) Metamorphic rocks are formed when existing rocks are changed by high temperatures or high _____. Marble is made from _____ and slate is made from _____.

d) Metamorphic rocks are quite hard and shiny, and some have bands of _____. Some low-grade metamorphic rocks may contain altered fossils, but the fossils are completely _____ in higher-grade metamorphic rocks.

e) Igneous rocks are formed when molten rock _____ down and solidifies. Molten rock below the surface of the Earth is called _____. It cools slowly to form _____ crystals.

f) Molten rock above the surface of the Earth is called _____. It cools more quickly to form _____ crystals.

2 Study the boxes below, then answer the questions that follow.

Type of Rock	Way it was Formed
Sedimentary	Grains of sediment are cemented together by dissolved salts
Metamorphic	Molten rock cools down and solidifies
Igneous	Existing rocks are changed by high temperatures and pressures

a) Link the type of rock to the way it was formed. Draw one line from each type of rock.
b) What is a fossil?
c) Which of these types of rock is most likely to contain fossils?
d) Which type of rock will never contain fossils? Explain your answer.

The Rock Cycle

Skills Practice

Robyn has samples of four different igneous rocks. They are labelled A, B, C and D.

She measures the size of the crystals in each sample.

1 Name the piece of apparatus that Robyn could use to measure the size of the crystals in each sample.

Robyn's results are in the table below.

Rock Sample	Crystal Size (mm)
A	1
B	6
C	4
D	2

2 Use the information in the table to complete the graph opposite.

3 Use the graph to suggest which of these igneous rocks formed most slowly. Explain your answer.

4 Complete the table below to show the factor that Robyn changed, the variable that she measured and the variable that she kept the same in her experiment. Place one tick in each row.

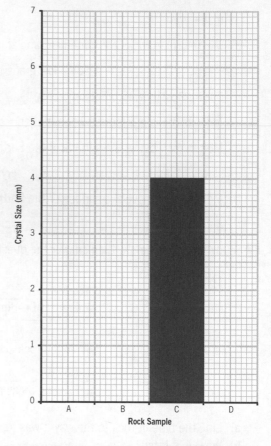

	Factor that was Changed	Factor that was Measured	Factor that was Kept the Same
Crystal Size			
Rock Sample			
Type of Rock that was Used			

 For further practice, go to p.44-47 of the Year 8 Science Essentials Workbook

Metals and Metal Compounds

Properties of Metals

Metals have special properties because of the way their atoms are arranged. These special properties make metals very useful, for example:

- Copper is a good thermal and electrical conductor, and is very resistant to corrosion. It's used to make water pipes and saucepans, and is used in electrical wiring.
- Aluminium has a low density and is very resistant to corrosion. It's used to make drinks cans and bicycle frames.

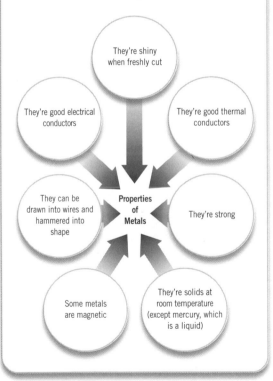

Properties of Non-Metals

These are some of the characteristic properties of non-metals:

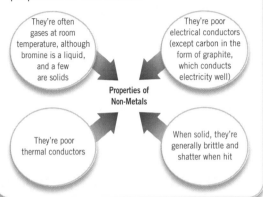

Symbols and Formulae

Every **element** can be represented by a simple one or two-letter code. For example, copper has the **symbol** Cu and aluminium has the symbol Al. A chemical **formula** shows the type and number of atoms in the smallest particle of a substance.

Three commonly used **acids** are hydrochloric acid (HCl), sulfuric acid (H_2SO_4) and nitric acid (HNO_3). The chemical formulae of these acids show that they contain the element **hydrogen** (H), which is true of all acids.

Chemical symbols allow scientists who speak different languages to communicate with each other. This allows them to share developments and understanding.

Evidence of a Chemical Reaction

In a chemical reaction, new substances are made. Evidence that a chemical reaction has taken place includes...

- a change in colour
- a change in temperature
- bubbles that show a gas is being made.

Metals and Metal Compounds

How Metals React with Acids

Some metals react with acids, forming a salt and releasing hydrogen gas. This equation allows you to predict what will happen in these reactions:

By analysing the reaction between different metals and an acid, you can place the metals into an order of reactivity (see page 156).

Copper is a very unreactive metal and doesn't react with acids, so no bubbles of hydrogen are seen.

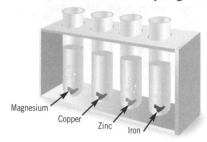

Magnesium   Copper   Zinc   Iron

The Reaction with Hydrochloric Acid

Metals react with hydrochloric acid to form chloride salts. Magnesium reacts with hydrochloric acid to form magnesium chloride and hydrogen gas. Lots of bubbles of hydrogen can be seen. Heat is also given out and the test tube becomes warmer.

magnesium + hydrochloric acid $\longrightarrow$ magnesium chloride + hydrogen

$$Mg + 2HCl \longrightarrow MgCl_2 + H_2$$

The Reaction with Sulfuric Acid

Metals react with sulfuric acid to form sulfate salts. Here is an example with zinc:

zinc + sulfuric acid $\longrightarrow$ zinc sulfate + hydrogen

$$Zn + H_2SO_4 \longrightarrow ZnSO_4 + H_2$$

When calcium reacts with sulfuric acid, an **insoluble** salt called calcium sulfate is formed. The salt forms an insoluble layer around the metal, so very little reaction is actually observed.

The Reaction with Nitric Acid

Metals react with nitric acid to form nitrate salts. Here is an example with zinc:

zinc + nitric acid $\longrightarrow$ zinc nitrate + hydrogen

$$Zn + 2HNO_3 \longrightarrow Zn(NO_3)_2 + H_2$$

Note the brackets followed by the number 2 in the formula for zinc nitrate. This means that everything inside the brackets is multiplied by two, so the smallest particle of zinc nitrate contains one zinc atom, two nitrogen atoms and six oxygen atoms.

How Metal Carbonates React with Acids

Metal carbonates...
- contain atoms of a metal, carbon and oxygen
- react with acids to form a salt, water and **carbon dioxide** gas
- form chloride salts when reacting with hydrochloric acid, sulfate salts when reacting with sulfuric acid, and nitrate salts when reacting with nitric acid.

| metal carbonate | + | acid | → | salt | + | water | + | carbon dioxide |

This is the reaction between calcium carbonate ($CaCO_3$) and hydrochloric acid:

| calcium carbonate | + | hydrochloric acid | → | calcium chloride | + | water | + | carbon dioxide |

$$CaCO_3 + 2HCl \longrightarrow CaCl_2 + H_2O + CO_2$$

Reacting metal carbonates with acids can be used to make salts of less reactive metals like copper, which don't react directly with acids.

Identifying Hydrogen and Carbon Dioxide

You can tell hydrogen or carbon dioxide gas has been produced in a reaction because...
- hydrogen is a flammable gas that burns with a squeaky pop
- carbon dioxide turns limewater cloudy.

Gas	Test	Result if Gas is Present
H_2	Place a lighted splint near by	Burns with a squeaky pop
CO_2	Bubble through limewater	Limewater turns cloudy

Testing for Hydrogen

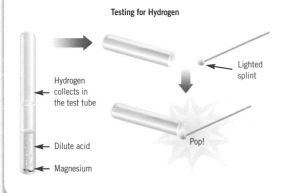

Hydrogen collects in the test tube

Lighted splint

Pop!

Dilute acid

Magnesium

Testing for Carbon Dioxide

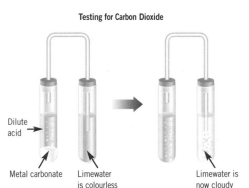

Dilute acid

Metal carbonate

Limewater is colourless

Limewater is now cloudy

How Metal Oxides React with Acids

Metal oxides...
- contain atoms of a metal and of oxygen
- react with acids to form a salt and water
- form chloride salts when reacting with hydrochloric acid, sulfate salts with sulfuric acid, and nitrate salts with nitric acid.

| metal oxide | + | acid | → | salt | + | water |

For example, copper oxide reacts with sulfuric acid to form the salt copper sulfate and water:

| copper oxide | + | sulfuric acid | → | copper sulfate | + | water |

$$CuO + H_2SO_4 \longrightarrow CuSO_4 + H_2O$$

Metals and Metal Compounds

Adding Acids to Alkalis

Acids and bases are chemical opposites. Metal carbonates, metal oxides and metal hydroxides are bases. Some metal oxides and hydroxides are soluble in water. Soluble bases are known as **alkalis**.

Two commonly used alkalis are...
- sodium hydroxide (NaOH)
- potassium hydroxide (KOH).

If exactly the right amount of acid is added to an alkali, a neutralisation reaction takes place and salt and water are made.

| alkali | + | acid | → | salt | + | water |

For example, hydrochloric acid reacts with the alkali sodium hydroxide to form the salt sodium chloride and water:

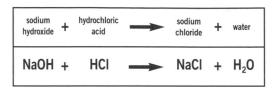

| sodium hydroxide | + | hydrochloric acid | → | sodium chloride | + | water |

$$NaOH + HCl \longrightarrow NaCl + H_2O$$

Again, the type of salt formed depends on whether hydrochloric acid, sulfuric acid or nitric acid is used.

Separating Soluble Salts

The salts formed in a reaction can be separated:
- First the solution is filtered to remove any unreacted metal or metal compound (because it's insoluble). This leaves the soluble salt and water.
- The solution is then warmed. The water **evaporates** to leave crystals of the salt. The slower the water evaporates, the larger the crystals will be.

Quick Test

1. Name the gas produced when a metal reacts with an acid.
2. Name the salt produced when zinc metal reacts with sulfuric acid.
3. Name the salt produced when sodium carbonate reacts with hydrochloric acid.
4. Name the salt produced when copper oxide reacts with nitric acid.
5. What is the general equation for the reaction between an alkali and an acid.

KEY WORDS
Make sure you understand these words before moving on!
- Acid
- Alkali
- Carbon dioxide
- Element
- Evaporates
- Formula
- Hydrogen
- Insoluble
- Metal carbonate
- Symbol

Key Words Exercise

Match each key word with its meaning.

Acid	A soluble base
Alkali	A chemical compound that contains metal, carbon and oxygen atoms
Carbon dioxide	Doesn't dissolve
Element	Turns from a liquid to a gas
Evaporates	The chemical opposite of bases; these compounds all contain the element hydrogen
Formula	A one or two-letter code used to represent an element
Hydrogen	A substance that's made of only one type of atom
Insoluble	A gas, produced when a metal carbonate reacts with an acid, which turns limewater cloudy
Metal carbonate	A way of representing the type and number of atoms in the smallest particle of a substance
Symbol	A gas, produced when a metal reacts with an acid, which burns with a squeaky pop

Comprehension

Read the passage about salt, then answer the following questions.

1. What is the chemical name for salt?

2. Name three foodstuffs that may contain surprisingly high levels of salt.

3. What is the recommended daily salt allowance for an adult?

4. How much salt should children under the age of 11 have in their diet?

5. How many deaths are caused by strokes in England each year?

6. Name four health problems that are linked to high blood pressure.

Sodium chloride is better known as table salt or simply 'salt'. Most of the salt you consume is already in the foods when you buy them. This means that the total amount of salt in your diet can be much higher than you think.

It's easy to guess that foods that taste salty, such as crisps, may contain quite high levels of salt. But other foods, such as some breakfast cereals, ready-made sandwiches and pasta sauces, contain surprisingly high levels of salt too.

The recommended daily allowance of salt for children over the age of 11 and adults is only 6g, and it's better to have less than this amount. Younger children should have less salt and babies should have less than 1g of salt a day. If children are allowed to consume larger amounts of salt, it can damage their health.

Eating too much salt can cause serious health problems. It can cause high blood pressure, which increases the risk of suffering a stroke. A stroke is a serious medical condition, which causes 50 000 deaths in England every year. Someone with high blood pressure is three times more likely to suffer a stroke. High blood pressure also increases your chance of suffering heart disease. High blood pressure can also cause damage to your eyes and kidneys.

Metals and Metal Compounds

Testing Understanding

1 **Fill in the missing words to complete the sentences about metals and metal compounds.**

a) Metals are _____ conductors of heat and electricity. Only one non-metal element, carbon in the form of _____, conducts electricity.

b) The chemical formula of a substance tells us the _____ and number of atoms present in the smallest particle of a substance. Sulfuric acid has the formula H_2SO_4. This means that the smallest particle of sulfuric acid contains _____ hydrogen atoms, _____ sulfur atom and _____ oxygen atoms.

c) Some metals react with acids to form a salt and a gas called _____. This gas can be identified because it burns with a _____ pop. Magnesium reacts with nitric acid to form a salt called magnesium _____ and hydrogen gas.

d) Metal carbonates react with acids to form a _____, water and carbon dioxide gas. Carbon dioxide can be identified because it turns _____ cloudy.

e) Metal oxides and hydroxides also react with acids. They form a salt and _____. Copper oxide reacts with hydrochloric acid to form a salt called copper _____ and water. No bubbles are seen during this reaction because no _____ is made.

2 **Answer the following questions about iron, which is a very useful metal.**

a) Complete the word equation below to show the reaction between iron and hydrochloric acid.

iron + hydrochloric acid ⟶ _____ + _____

b) When iron carbonate reacts with hydrochloric acid, a gas called carbon dioxide is produced. How could you prove that this gas is really carbon dioxide?

c) Iron can be made into stainless steel. Stainless steel is an alloy (a mixture of metals). A sample of stainless steel contains 70% iron, 10% nickel and 20% chromium. Copy and complete the pie chart below to show the composition of this sample of stainless steel. Remember to add labels.

Skills Practice

George wants to investigate the factors affecting rusting. The table below describes what the substances in the test tubes do.

Substance	What it Does
Calcium chloride	Removes water
Boiled water	The water doesn't contain oxygen
Layer of oil	Stops oxygen entering the water

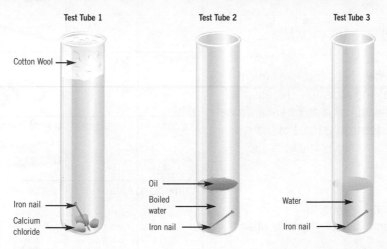

Test Tube 1   Test Tube 2   Test Tube 3

Cotton Wool

Iron nail
Calcium chloride

Oil
Boiled water
Iron nail

Water
Iron nail

1 Copy and complete the table below by writing **yes** or **no** in each box to show the conditions experienced by the nails inside each test tube.

Test Tube	Any Water?	Any Oxygen?
1		
2		
3		

George leaves the nails in the test tubes for two weeks and then examines them. His results are in the table below.

Test Tube	Appearance of the Nail
1	No rust
2	No rust
3	Nail is very rusty

2 Write a conclusion for George's experiment.

3 George repeats his experiment but this time he uses iron nails that have been coated in oil. Describe how you would expect the nail in test tube 3 to look. Explain your answer.

The Reactivity of Metals

The Changing Appearance of Metals

Over time, many metals are affected by air and water. Different metals are affected in different ways and some examples are shown in the table opposite.

Gold objects don't react at all; gold is described as being unreactive.

Most metals are hard but a few, like lithium (Li), sodium (Na) and potassium (K), are much softer and can be cut with a sharp knife. These three metals are found in group 1 of the periodic table. This group is known as the alkali metals.

Alkali metals...
- are shiny when freshly cut
- **tarnish** (become dull) very quickly when exposed to air.

Metal	Example Object	How the Object Changes Over Time
Iron	Nail	It **rusts**
Silver	Ring	It becomes dull
Copper	Water pipe	It gets darker
Aluminium	Aluminium can	It goes grey

The Reactivity Series

You can place metals into an order of reactivity by analysing how they react with water, oxygen or acid. This enables you to arrange metals from the most reactive to the least reactive in a **reactivity series**.

The order of reactivity for the metals is the same whether water, oxygen or acid is used, and whichever acid is used.

The diagram opposite shows a sample of metal placed in acid. The rate of reaction can be judged by...
- counting the number of bubbles of gas released
- measuring the temperature change using a thermometer.

Reactions of Metals with Water

Some metals react with cold water to form metal hydroxides (or metal oxides) and hydrogen gas. Metal hydroxide solutions are **alkaline** and have a pH above 7.

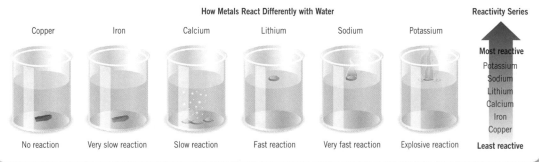

Lithium, sodium and potassium all float on water. This shows they're less dense than water. These metals react **vigorously** (quickly) with water:

- The hydrogen released when sodium reacts with water can be lit using a lighted splint.
- Potassium reacts so vigorously that the hydrogen released burns readily.
- Lithium, sodium and potassium are all stored in oil to stop them reacting with oxygen or moisture in the air.

Sodium is one of the metals that reacts with cold water:

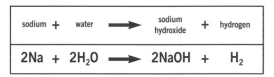

The experiment above can be carried out safely in the laboratory if...

- the teacher carries out the experiment as a demonstration
- the teacher wears goggles
- only very small pieces of metal are used
- a safety screen is used to protect people observing the reaction.

Copper doesn't react with water, while iron and calcium only react slowly.

How Metals React Differently with Water

Copper	Iron	Calcium	Lithium	Sodium	Potassium
No reaction	Very slow reaction	Slow reaction	Fast reaction	Very fast reaction	Explosive reaction

Reactivity Series

Most reactive
Potassium
Sodium
Lithium
Calcium
Iron
Copper
Least reactive

Reactions of Metals with Acids

You'll remember that some metals react with acids to form a salt and hydrogen:

The more reactive the metal, the more bubbles are seen and the higher the temperature rise. The same amount of acid and the same amount of metal should be used to make this a fair test.

How Metals React Differently with Acids

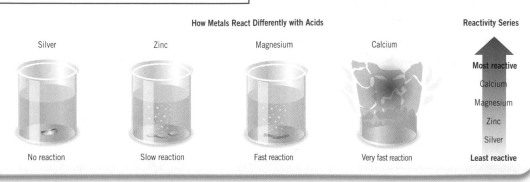

Silver	Zinc	Magnesium	Calcium
No reaction	Slow reaction	Fast reaction	Very fast reaction

Reactivity Series

Most reactive
Calcium
Magnesium
Zinc
Silver
Least reactive

The Reactivity of Metals

Reactions of Metals with Oxygen

Some metals react with oxygen to form metal oxides:

Metals react with oxygen in the following ways:
- When metals are burned in air, they react with the oxygen.
- If a metal is burned in pure oxygen, the reaction is much more vigorous.
- The higher up the reactivity series a metal is, the more vigorously it will be expected to react with oxygen.

- Heating a metal using a Bunsen burner speeds up the rate at which it reacts with oxygen.

It's very important to work safely in the laboratory. When using a Bunsen burner to heat the metals...
- long hair should be tied back
- goggles must be worn
- tongs must be used to hold the metal
- avoid looking directly at burning magnesium.

How Metals React Differently with Oxygen

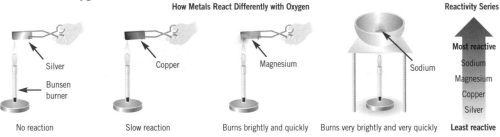

Displacement Reactions

A more reactive metal **displaces** a less reactive metal from its compound. Iron is more reactive than copper, so iron displaces copper from a solution of copper sulfate.

You can use the reactivity series to predict displacement reactions. Zinc is more reactive than copper, so you can predict that zinc will displace copper from a solution of copper sulfate.

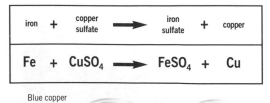

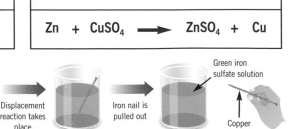

Useful Displacement Reactions

Displacement reactions can be very useful in everyday life. For example, broken iron railway tracks can be mended using **molten** (liquid) iron. The iron is produced using the displacement reaction between iron (III) oxide and aluminium:

- Aluminium is more reactive than iron, so it displaces iron from iron oxide.
- So much heat is released by the reaction that the iron is molten, which means it can be poured directly into the gaps in the tracks.
- This is known as a thermite reaction.

Extracting Metals

Copper and gold were the first metals known and used by people. They're very unreactive:

- Gold is found uncombined in nature.
- Most copper is found in compounds in rocks; it can be easily extracted from these compounds by heating the rocks in the presence of carbon.

Very reactive metals, like sodium and potassium, are only found in compounds. These metals…

- are difficult to extract from their compounds; normally they're extracted using electricity
- were discovered much more recently.

Using Metals

The way that a metal is used depends on its properties and how reactive it is.
Copper is used…

- to make water pipes because it has low reactivity
- in electrical wiring because it's a good **electrical conductor**
- to make saucepans because it's a good **thermal conductor**.

Aluminium has a low density and is much less reactive than might be expected because it reacts with oxygen to form a layer of aluminium oxide, which stops any further reaction.
Aluminium is used to…

- form aluminium alloys, which are used to make aircraft because they have a low density
- make greenhouse frames because the layer of aluminium oxide stops any further reaction.

Quick Test

1. Name a metal that doesn't tarnish, even over very long periods of time.
2. Which solution is alkaline: solution A (pH 7); solution B (pH 3); solution C (pH 10)?
3. Name the compound formed when copper reacts with oxygen.
4. Complete this equation:
   copper sulfate + iron ⟶ _____ + _____

KEY WORDS
Make sure you understand these words before moving on!
- Alkaline
- Displaces
- Electrical conductor
- Molten
- Reactivity series
- Rust
- Tarnish
- Thermal conductor
- Vigorously

The Reactivity of Metals

Key Words Exercise

Match each key word with its meaning.

Alkaline	A material that electricity can pass through easily
Displaces	A solution that has a pH greater than 7
Electrical conductor	Quickly
Molten	A list that places the metals in order from the most reactive to the least reactive
Reactivity series	A compound formed when iron reacts with oxygen and water
Rust	Liquid
Tarnish	Become dull
Thermal conductor	A material that heat can pass through easily
Vigorously	Takes the place of

Comprehension

Read the passage about rusting, then answer the following questions.

1. What is an alloy?

2. Why do different steels have different properties?

3. **a)** Why is stainless steel a very useful alloy?
   b) How is stainless steel used?

4. What is the chemical name for rust?

5. Why does a layer of paint prevent iron from rusting?

Iron is usually turned into an alloy. Alloys are mixtures of metals. The most common alloy of iron is called steel. Different types of steel can be made by using different types of metals in different proportions. The different steels have different properties.

The type of steel selected for a particular application will depend on the properties required. Stainless steel is made by mixing iron and chromium. It's very resistant to corrosion and is widely used to make cutlery and saucepans.

If iron or iron alloys are exposed to water and oxygen, they will eventually react to form hydrated iron oxide (rust). Even stainless steel will eventually rust. Many other metals also react to form metal oxides, but these metal oxides aren't referred to as rust.

Iron can be protected from rusting by coating the metal with a layer of plastic, paint or oil. This stops the oxygen and water from reaching the metal. But if the coating is scratched, the iron will start to rust. Iron can also be protected from rusting by placing it in contact with a more reactive metal, such as magnesium. The more reactive metal reacts, leaving the iron intact.

Testing Understanding

1 **Fill in the missing words to complete the sentences about the reactivity of metals.**

a) Over time, the appearance of many metals changes as they're affected by

_____ and _____. Silver objects become

_____ and iron objects _____. But gold objects remain

the same because gold is a very _____ metal.

b) Magnesium is a fairly reactive metal. It reacts with hydrochloric acid to form a salt called

_____ and hydrogen gas.

c) When metals are heated, they react with _____ in the air to form metal

_____. By observing the reaction between metals and oxygen, you can place

the metals into an order of reactivity from the most to the _____ reactive.

d) In displacement reactions, a _____ reactive metal displaces a

_____ reactive metal from its compound. In the reaction between iron

and copper sulfate, the iron, which is _____ reactive, displaces the

copper to form _____ and _____.

2 **Read the information provided, then answer the questions that follow.**

Andy wanted to investigate what happened when different metals were placed in dilute acid. He measured 10cm^3 of acid and then placed it into a boiling tube. He measured the temperature of the acid. He then placed a piece of zinc into the acid and measured the new temperature.

Metal Used	Temp. at Start (°C)	Temp. at End (°C)	Temp. Change (°C)
Zinc	21	23	
Copper	21	21	
Magnesium	20	23	

Andy repeated the experiment using copper and magnesium metals.

a) Name the piece of equipment that Andy could use to measure the temperature.

b) Name the piece of equipment that Andy could use to measure the volume of acid.

c) Copy and complete the table above right to show the temperature change for each metal.

d) Copy and complete the axes opposite, and show the results of Andy's experiment as a bar graph.

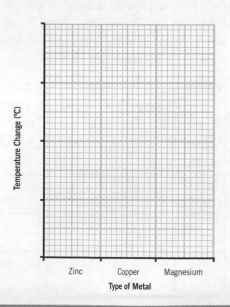

Temperature Change (°C)

Zinc   Copper   Magnesium

Type of Metal

The Reactivity of Metals

Skills Practice

Pauline wants to investigate the reactivity of different metals by measuring the temperature rise when they're placed in dilute acid. Her experiment is illustrated below.

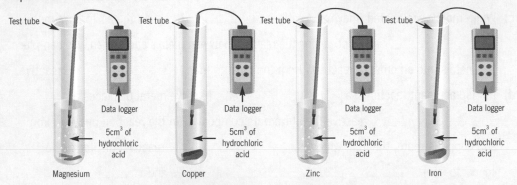

1 a) Pauline uses a data logger to measure the temperature rise in her experiments. Suggest another piece of apparatus that Pauline could use to measure the temperature change.

   b) Suggest one advantage of using a data logger.

2 Copy and complete the table below to show the factors that Pauline has changed, measured and kept the same in her experiment. Put one tick in each row.

Factor	Is it Changed?	Is it Measured?	Is it Kept the Same?
Volume of acid			
Type of metal			
Type of acid			
Temp. rise			

Pauline's results are in the table below.

Type of Metal	Temperature Rise (°C)
Magnesium	5
Copper	0
Zinc	2
Iron	1

3 Copy the axes below and show the results of Pauline's experiment as a bar graph.

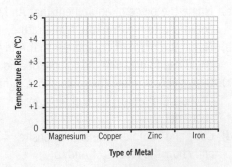

4 Put the metals in this experiment into an order of reactivity. Place the most reactive metal first.

For further practice, go to p.20-23 of the Year 9 Science Essentials Workbook

Environmental Chemistry

The Changing Environment

The environment is changed by...
- human activities
- natural processes.

Scientists around the world work together to monitor these changes and to develop ideas that have changed the way that people think and behave.

What is in Soil?

Soils contain...
- **weathered** pieces of rock of different shapes and sizes
- different amounts of water
- bacteria and **fungi**, which decompose organic material and release minerals that can be used by new plants
- animals, such as earthworms and moles. Earthworms improve the soil by aerating it and by dragging organic material, such as leaves, down from the surface of the soil.

Soils also contain a dark, sticky material called **humus**. Humus...
- is the remains of dead plants and animals that have rotted away
- contains minerals that help new plants to grow
- helps to stick the rock particles together and helps to hold water.

Plants grow best in soils that have high levels of humus.

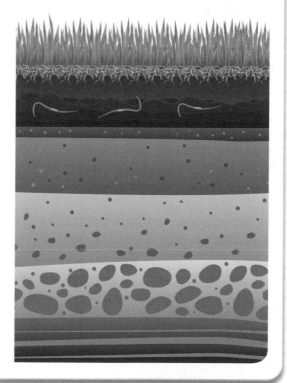

Satellites

Satellites can be used to take photographs of the Earth's surface. Some plants grow particularly well in certain types of soil.

Scientists can use these satellite photographs to...
- monitor which plants are growing
- identify the type of soil and the underlying rock.

Environmental Chemistry

The pH of Soils

Most soils have a **pH** range of 4 to 8. Soils...
- in limestone areas typically have a pH of about 8
- that have high levels of humus or are in areas affected by **acid rain** tend to be slightly acidic.

Most crops only grow well in soils that have a certain pH range.

Plant	Preferred pH Range
Potato	4.5–6.0
Carrot	5.5–7.0
Pea	6.0–7.5

If a soil becomes too...
- acidic, farmers can add powdered lime (calcium carbonate) or **quicklime** (calcium hydroxide)
- alkaline, farmers can add manure. As the manure rots down, it produces acids that reduce the pH of the soil.

Weathering

Over time, rocks are broken into smaller pieces.

Chalk, limestone and marble all contain the chemical compound calcium carbonate ($CaCO_3$). These rocks, and building materials made from these rocks, are chemically weathered.

In areas of high rainfall and where air pollution has produced acid rain, the rocks are weathered more quickly.

Acid Rain

The atmosphere contains a small amount of carbon dioxide. It's produced...
- when animals and plants respire
- by volcanic activity
- by the burning of **fossil fuels**, such as coal, oil and natural gas, which contain carbon.

Carbon dioxide dissolves in moisture to produce weakly acidic rainwater. Normal rainwater has a pH of about 5.6.

carbon dioxide + water ➡ carbonic acid

Sulfur dioxide and nitrogen oxides lower the pH of rainwater even more to form acid rain, which typically has a pH of 3 to 5.

Sulfur dioxide is a toxic and corrosive gas that's produced by...
- volcanic activity
- the burning of fossil fuels, many of which contain traces of sulfur.

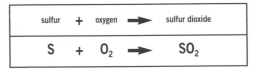

sulfur + oxygen ➡ sulfur dioxide

$$S + O_2 \rightarrow SO_2$$

The sulfur dioxide dissolves in rainwater to form acid rain or, if it's very cold, acid snow. Acid rain affects wildlife, rocks and metals.

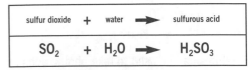

sulfur dioxide + water ➡ sulfurous acid

$$SO_2 + H_2O \rightarrow H_2SO_3$$

Nitrogen oxides (NO_x) are produced by...
- lightning strikes
- the burning of fossil fuels
- the high temperatures produced by car exhausts.

The Effects of Acid Rain

Acid rain can…
- kill trees
- kill plants and animals living in affected lakes
- dissolve rocks that contain metal carbonates (for example, limestone, marble and chalk, which contain calcium carbonate)
- damage statues and buildings made from rocks that contain metal carbonates
- corrode metals such as iron.

In lakes, the eggs of fish can also be damaged as the acidity increases. Young fish can be born deformed or may not hatch at all, while other organisms may grow more quickly.

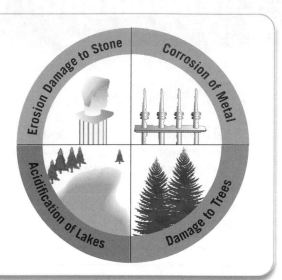

Reducing the Effects of Acid Rain

Neutralisation reactions can reduce the effects of acid rain:
- Powdered lime can be added to affected lakes. Lime neutralises the acidic compounds in the water.
- Acidic soils can be neutralised by adding powdered limestone (calcium carbonate) or quicklime (calcium hydroxide), which neutralise the acidic compounds in the soil.

When fossil fuels are burned, the amount of sulfur dioxide released into the atmosphere can be reduced by…
- removing sulfur from oil and natural gas before they're burned
- installing 'scrubbers' in power stations to remove sulfur dioxide from the exhaust gases produced when coal is burned
- using low-sulfur coals.

Atmospheric Pollution

Carbon dioxide is produced when all living things respire, and it's taken in when plants photosynthesise. But this balance has been upset by human activities, such as…
- people burning more fossil fuels
- forests being cut down for timber, for mining and to build new roads or homes.

Deforestation means that there are fewer trees left to photosynthesise, so less carbon dioxide is removed from the atmosphere. Some people remove trees by burning them, which increases the carbon dioxide in the atmosphere.

Scientists monitor the levels of pollutant gases in the atmosphere. The amount of carbon dioxide is so low that it's measured in 'parts per million' (ppm), but the increase of this gas in the atmosphere is a concern.

Scientists are particularly concerned about the loss of trees in tropical rainforests. They believe that some species may become extinct and there will be less variation in the species that survive.

Environmental Chemistry

Global Warming

Scientists believe that the Earth is warming up. Their evidence includes...
- an increase in the average global temperature
- some areas becoming wetter and others becoming drier, leading to enormous disruption to agriculture
- a decrease in the amount of Arctic sea ice
- a rise in sea levels as ice melts and seawater expands as it's warmed
- an increase in extreme weather events around the world (e.g. flooding and hurricanes).

The Greenhouse Effect

The **greenhouse effect** is causing global warming. A greenhouse effect occurs naturally on Earth. Without it, the Earth would be much cooler and life may not have developed:
- A layer of carbon dioxide and other greenhouse gases, such as methane, act as a blanket around the Earth.
- Radiation from the Sun warms the surface of the Earth. As the heat energy is reflected into space, some is trapped by the carbon dioxide in the atmosphere, so the Earth warms up.
- The more carbon dioxide in the atmosphere, the more heat energy is trapped and the warmer the Earth becomes.

Scientists believe that human activities, such as the burning of fossil fuels, are increasing the warming of the Earth at a dramatic rate.

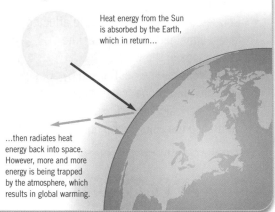

Heat energy from the Sun is absorbed by the Earth, which in return...

...then radiates heat energy back into space. However, more and more energy is being trapped by the atmosphere, which results in global warming.

Quick Test

1. Humus is found in soil. What is humus?
2. Suggest how a farmer could reduce the pH level of an alkaline soil.
3. Why is rainwater naturally acidic?
4. How is acid rain formed?
5. How are nitrogen oxides produced naturally?
6. Name two greenhouse gases.

Key Words Exercise

Match each key word with its meaning.

Acid rain	A chemical, formed when carbon dioxide dissolves in water, which makes normal rainwater slightly acidic
Carbonic acid	Calcium hydroxide
Deforestation	The cutting down of large numbers of trees
Fossil fuels	The reaction between an acid and an alkali
Fungi	How gases, including carbon dioxide and methane, cause the Earth to warm up
Greenhouse effect	A toxic gas formed when sulfur is burned
Humus	A type of microbe that decomposes plant and animal material
Neutralisation	A scale used to measure how acidic or alkaline something is
pH	When a rock is broken down into smaller pieces
Quicklime	An object that orbits a larger body
Satellite	A type of rain formed in polluted areas when sulfur oxides or nitrogen oxides dissolve in water
Sulfur dioxide	Non-renewable energy resources, such as coal, oil and natural gas, formed from plants and animals that lived long ago
Weathered	Dark, sticky material formed when plant and animal matter rots

Comprehension

Read the passage about lichens, then answer the following questions.

1. Name two types of animal that live in lichens.

2. Why are lichens useful to people?

3. What is a symbiotic relationship?

4. What benefits does each plant bring to their relationship?

5. If the number of types of lichen growing in a city decreases, what does this indicate?

Lichens are commonly found on trees, rocks and walls. They come in a wide range of colours and sizes, and provide homes for spiders and insects. People use them in the manufacture of some sunscreens and antibiotics.

Lichens aren't single plants. Each lichen is made up of two plants that live together. The plants have a symbiotic relationship; each plant benefits from the presence of the other. One plant is a fungus and the other is a green alga. The alga can photosynthesise to provide food for both plants, while the fungus creates a 'thallus' that houses both plants.

Lichens are very useful as environmental or bio indicators. They're very sensitive to air pollution. Generally, the smaller the number of species of lichen growing in an area, the more polluted the air is. Lichens are very sensitive to sulfur dioxide pollution in the air. Since the levels of sulfur dioxide pollution have fallen over the last 30 years, the number of species of lichen growing in urban areas has greatly increased.

Environmental Chemistry

Testing Understanding

1 **Fill in the missing words to complete the sentences about environmental chemistry.**

a) Most soils have a _____ between 4 and 8. Soils in limestone areas are

slightly _____ and have a pH of around 8, while soils in polluted areas may

be slightly _____ and have a lower pH. If a soil is too acidic, farmers can add

calcium carbonate (lime) or calcium hydroxide (_____) to increase its pH. If a

soil is too _____, farmers can add manure to the soil. As the manure rots, it

produces acidic compounds that _____ the pH of the soil.

b) Normal rainwater is slightly acidic because the gas, _____ dioxide,

dissolves in the water to form carbonic acid. In polluted areas, sulfur dioxide may also dissolve

in the rainwater to form _____ rain. Acid rain can also be produced by

_____ oxides. Acid rain can damage buildings made from rocks that

contain the chemical compound calcium carbonate, such as limestone, chalk or

_____.

c) The greenhouse effect is causing _____ warming. A layer of greenhouse

gases, including _____ dioxide and methane, acts as a blanket around

the Earth. This layer stops some heat _____ from escaping into space, so

the planet warms up.

2 **Read the information about carbon dioxide in the atmosphere, then answer the questions that follow.**

The amount of carbon dioxide in the atmosphere is changing. The table below shows the
amount of carbon dioxide gas in the atmosphere at different times.

a) What does 'ppm' mean?

b) Copy the axes opposite and plot the results as a line graph.

c) What is happening to the level of carbon dioxide in the atmosphere?

d) Suggest why the level of carbon dioxide in the atmosphere is changing.

Year	Concentration of Carbon Dioxide (ppm)
1960	315
1970	328
1980	342
1990	356
2000	369

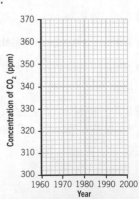

Skills Practice

A scientist wanted to know if the level of carbon dioxide in the air had changed over time.

Every five years, she returned to the same road and measured the level of carbon dioxide in the air. The table below shows her results.

Year	Level of Carbon Dioxide (ppm)
1980	336
1985	344
1990	350
1995	357
2000	368
2005	370

1 a) Use the results in the table to draw a line graph. Remember to label both axes and plot the results carefully.

  b) Add a line of best fit.

  c) Circle the anomalous result and suggest why this result might be anomalous.

2 Predict what the level of carbon dioxide will be in 2020.

3 A second scientist suggests monitoring the levels of carbon dioxide in the road every six months.

  a) Give a possible advantage of the second scientist's method.

  b) Give a possible disadvantage of the second scientist's method.

Using Chemistry

Conservation of Mass

The chemicals...
- that start a reaction are the reactants
- made by the reaction are the products.

During chemical reactions...
- new materials are made, but the total mass of the reactants is equal to the mass of the products
- no atoms are made or destroyed, they're just rearranged, so you say the mass is 'conserved'.

In the example below, by counting the different types of atom on each side of the equation, you can see that the mass of the reactants equals the mass of the products.

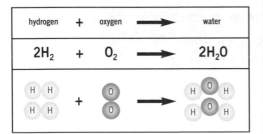

hydrogen	+	oxygen	→	water
$2H_2$	+	O_2	→	$2H_2O$

You'll see further examples of conservation of mass throughout this topic.

Burning Fuels

A **fuel** can be burned to release heat energy. Many fuels contain only carbon and hydrogen. These fuels are called **hydrocarbons**.

Three Examples of Fuels

Paraffin oil   Candle wax   Wood

Complete Combustion

Combustion is an example of an oxidation reaction. **Complete combustion** occurs when a hydrocarbon fuel is burned in a good supply of oxygen. When this happens...
- the carbon reacts with oxygen to form carbon dioxide
- the hydrogen reacts with oxygen to form water vapour.

This is the equation for the complete combustion of methane, the hydrocarbon fuel used by Bunsen burners:

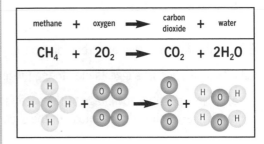

methane	+	oxygen	→	carbon dioxide	+	water
CH_4	+	$2O_2$	→	CO_2	+	$2H_2O$

Candle wax is also a hydrocarbon fuel. The apparatus below can be used to test the gases produced when a candle is burned:
- The water vapour condenses to form a colourless liquid.
- The carbon dioxide turns the **limewater** cloudy.

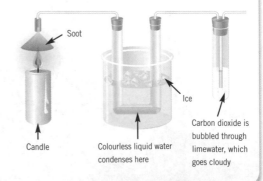

Soot

Candle

Colourless liquid water condenses here

Ice

Carbon dioxide is bubbled through limewater, which goes cloudy

Incomplete Combustion

Incomplete combustion occurs when a fuel is burned in a limited supply of oxygen. When this happens…

- there isn't enough oxygen available for the fuel to burn completely
- less heat energy is released
- the flame is yellow because it contains unburned carbon, which can be deposited on surfaces as soot
- a poisonous gas called carbon monoxide is produced. Carbon monoxide combines with the haemoglobin in red blood cells and stops them from being able to carry oxygen.

This is the equation for the incomplete combustion of methane:

methane	+	oxygen	$\longrightarrow$	carbon monoxide	+	water

$$2CH_4 + 3O_2 \longrightarrow 2CO + 4H_2O$$

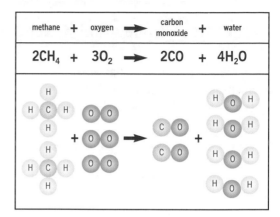

Using Hydrogen as a Fuel

When hydrogen is burned…

- it reacts with oxygen to form water vapour
- lots of heat energy is released.

hydrogen	+	oxygen	$\longrightarrow$	water

$$2H_2 + O_2 \longrightarrow 2H_2O$$

When hydrogen is burned, only water vapour is made. As the fuel doesn't contain carbon, carbon dioxide isn't produced. So, burning hydrogen doesn't add to the greenhouse effect.

Hydrogen is used as a fuel in rockets. As there's no air in space, rockets have to carry supplies of liquid hydrogen and liquid oxygen.

Matches

Matches can be used to light fires:

- A match head contains the elements carbon and sulfur, and the compound potassium chlorate ($KClO_3$). Compounds with names that end in 'ate' contain oxygen.
- Carbon and sulfur are fuels, and potassium chlorate is an oxidising agent. As the match head burns, the potassium chlorate releases oxygen that allows the two fuels to burn even better.

Self-Heating Cans

Climbers and explorers can now enjoy warm drinks straight from a can. These cans contain two sections:

- The top section contains the drink, for example coffee.
- The bottom section contains two chemicals that are kept separate by a thin layer of foil.
- When the thin foil is broken, the two chemicals react together. Heat energy is released and the coffee is warmed up.

Using Chemistry

Other Useful Chemical Reactions

You'll remember that in **displacement reactions**, a more reactive metal takes the place of a less reactive metal, for example:

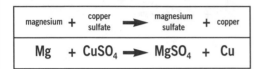

magnesium + copper sulfate ➡ magnesium sulfate + copper

$$Mg + CuSO_4 \rightarrow MgSO_4 + Cu$$

Magnesium ribbon

Copper sulfate solution

Copper coating

Magnesium sulfate

Some of the chemical energy in the reactants is transferred into heat energy. As a result, the temperature of the solution increases.

In fact, the greater the difference in reactivity between the two metals in a displacement reaction, the more heat energy is released.

This energy can also be released as electrical energy. The voltage of a cell gives an indication of the energy released.

You can measure the voltage using a voltmeter:
- The greater the difference in reactivity between the two metals, the more energy is released, so the higher the voltage measured.
- If two pieces of the same metal are used, there's no difference in reactivity and no voltage is produced.

New Materials Made by Chemical Reactions

A huge range of materials are made by chemical reactions. Many of these materials are natural. For example, plants produce glucose by **photosynthesis**:

carbon dioxide + water ➡ glucose + oxygen

However, some materials are made in factories. These are known as **synthetic materials** and many have very special uses.

New materials can benefit people enormously, but they must be carefully tested. For example, when a new medicine is developed, many years of tests are required in order to...
- check it's effective
- make sure it's safe to use and doesn't have dangerous side-effects
- work out the correct dose to treat patients.

Examples of Synthetic Materials

Shampoos – to remove dirt

Plastics – to make hundreds of things

Paint – to decorate and protect

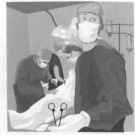

Anaesthetics – for painless operations

Zinc metal reacts with dilute hydrochloric acid:

zinc	+	hydrochloric acid	→	zinc chloride	+	hydrogen

$$Zn + 2HCl \longrightarrow ZnCl_2 + H_2$$

Bubbles of hydrogen gas

Hydrochloric acid

Zinc

The bubbles show that a gas is being made. As this gas escapes into the air, the mass of the test tube goes down. But the total mass of the products is the same as the total mass of the reactants, even when a gas is made.

When magnesium is burned in air, the metal combines with oxygen to form the compound magnesium oxide:

magnesium	+	oxygen	→	magnesium oxide

$$2Mg + O_2 \longrightarrow 2MgO$$

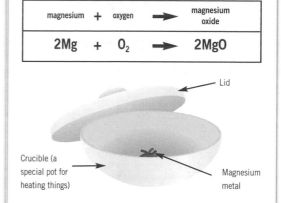

Lid

Crucible (a special pot for heating things)

Magnesium metal

The mass of the magnesium goes up because the magnesium has combined with oxygen, which has mass. However, the total mass of the reactants is still equal to the total mass of the products.

Burning Bread

Bread contains carbon and hydrogen. When bread is burned, the mass appears to go down as the carbon reacts with oxygen to form carbon dioxide gas and the hydrogen reacts with oxygen to form water vapour.

As the gases escape, the mass of the bread goes down but the overall mass is still conserved.

Quick Test

1. Define the term 'fuel'.
2. What are the products of the complete combustion of a hydrocarbon fuel?
3. Name the product of the combustion of hydrogen.
4. Where are synthetic materials made?
5. Explain what 'conservation of mass' means.
6. Why does the mass of magnesium go up when it's burned?

KEY WORDS

Make sure you understand these words before moving on!

- Complete combustion
- Displacement reaction
- Fuel
- Hydrocarbon
- Incomplete combustion
- Limewater
- Photosynthesis
- Synthetic material

Using Chemistry

Key Words Exercise

Match each key word to its meaning.

Complete combustion	When a material is burned in a limited supply of oxygen
Displacement reaction	A reaction in which a more reactive metal takes the place of a less reactive metal
Fuel	A material that's made in a factory
Hydrocarbon	A solution of calcium hydroxide that turns cloudy if carbon dioxide gas is bubbled through it
Incomplete combustion	The process by which green plants make glucose
Limewater	A substance that can be burned to release heat energy
Photosynthesis	When a material is completely burned in a good supply of oxygen
Synthetic material	A compound that contains only carbon and hydrogen

Comprehension

Read the passage about Joseph Priestley, then answer the following questions.

1. When was Joseph Priestley born?
2. Why was Joseph adopted?
3. Why is oxygen an important gas?
4. How is carbon dioxide used?
5. Why did Joseph have to leave his home?
6. What did Joseph do in America?

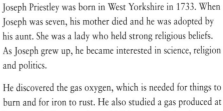

Joseph Priestley was born in West Yorkshire in 1733. When Joseph was seven, his mother died and he was adopted by his aunt. She was a lady who held strong religious beliefs. As Joseph grew up, he became interested in science, religion and politics.

He discovered the gas oxygen, which is needed for things to burn and for iron to rust. He also studied a gas produced at breweries by fermentation, which is now called carbon dioxide. Joseph found a way of dissolving the carbon dioxide into water. His discovery of carbonated water is widely used today in the manufacture of fizzy drinks. Joseph also discovered ammonia, sulfur dioxide and dinitrogen oxide or 'laughing gas'.

But when Joseph expressed support for the revolution that was happening in France, an angry mob attacked his home in Birmingham. Both his home and laboratory were destroyed. Joseph and his family fled to London, then to America. In America, Joseph established a church and continued his scientific work to improve people's lives.

Testing Understanding

1 **Fill in the missing words to complete the sentences about using chemistry.**

a) A _____ is a substance that can be burned to release heat energy. Many fuels contain only hydrogen and carbon and they're called _____.

b) If a hydrocarbon fuel is burned in a limited supply of oxygen, _____ combustion occurs and carbon dioxide, carbon _____ and soot are produced.

c) When magnesium is added to copper sulfate, a displacement reaction takes place and _____ and _____ are produced. During the reaction, _____ energy is released and the _____ of the solution increases. The greater the difference in reactivity between the two metals, the more _____ is released by the reaction.

d) During chemical reactions, new materials are _____ but the total mass of the _____ is equal to the total mass of the _____. Atoms aren't created or destroyed, they are simply _____.

2 **Read the information, then answer the questions that follow.**

Marianne carried out an investigation to find out how the mass of magnesium changes when it's burned. Her results are in the table below.

Experiment	Mass of Magnesium (g)	Mass of Magnesium Oxide (g)
1	0.50	0.82
2	0.30	0.50
3	0.10	0.16
4	0.26	0.32
5	0.16	0.26
6	0.42	0.69

a) Draw a graph to show Marianne's results. Remember to label each axis and include a line of best fit.

b) Describe the relationship between the mass of magnesium burned and the mass of magnesium oxide made.

c) One set of results doesn't fit the pattern. Circle this result on the graph and suggest why this result might be lower than expected.

Using Chemistry

Skills Practice

Sarah is investigating how she can make an electrical cell from half a lemon and two pieces of metal. She connects the strips of metal to a voltmeter, which is used to indicate the energy produced by the electrical cell.

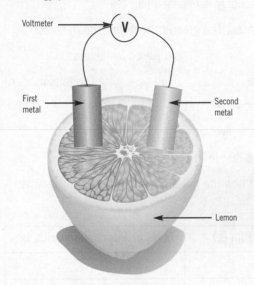

Sarah used strips of magnesium, copper and iron in her experiment. Two different metals were used each time.

1 Give the symbol used for each of these metals.
  a) The symbol for magnesium is ....
  b) The symbol for copper is ....
  c) The symbol for iron is ....

Sarah's results are shown in the table below.

2 Copy and complete the table by adding headings for each column.

Iron and copper	0.78
Magnesium and copper	2.71
Iron and magnesium	1.93

3 Copy and complete the table below to show the variables that Sarah changed, measured and kept the same in her experiment.

Variable Changed	
Variable Measured	
Variable Kept the Same	

4 Draw a bar graph to display the results from Sarah's experiment.

The diagram below shows a section of the reactivity series.

5 a) Which combination of different metals would produce the highest voltage?
  b) Which combination of different metals would produce the lowest voltage?
  c) Use the section of the reactivity series to suggest why these particular combinations of metals would produce the highest and lowest voltages.

Physics

Finding Energy

Energy

James Joule was the first scientist to realise that there are many different types of energy, for example...

- light energy
- sound energy
- kinetic (movement) energy
- gravitational potential energy
- thermal (heat) energy
- chemical energy
- electrical energy.

Energy is measured in units called Joules, named after James Joule.

1kJ or 1 kilojoule is equal to 1000J.

Energy Changers

Energy can't be created or destroyed, but it can be changed from one form to another, or transferred from one place to another.

Energy changers are devices that can change or transfer energy. Some energy-changing devices are given in this table:

Device	Energy Change
Kettle	Electrical energy into thermal energy. For example, it takes 500kJ to boil a kettleful of water.
Car engine	Chemical energy (in fuel) into kinetic energy. For example, it takes 100kJ to accelerate to 30mph.
Human	Chemical energy (in food) into kinetic energy (to move) and thermal energy (to keep warm). The average person uses 11 000kJ a day.
Microphone	Sound energy into electrical energy.
Loudspeaker	Electrical energy into sound energy.
Torch	Chemical energy (in batteries) into light and thermal energy. (The thermal energy heats up the bulb. It's not useful, so it's called wasted energy.)
Diver	Gravitational potential energy into kinetic energy. As a diver falls from a diving board, his speed increases as more energy is changed into kinetic energy.

Power Stations

Most of the energy you use is in the form of electricity. Electricity is produced by thermal power stations in the following way:

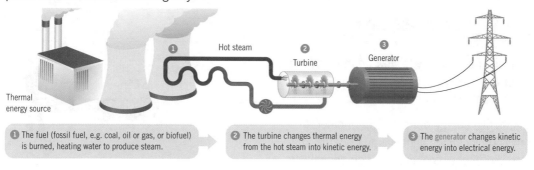

Hot steam

② Turbine

③ Generator

Thermal energy source

❶ The fuel (fossil fuel, e.g. coal, oil or gas, or biofuel) is burned, heating water to produce steam.

❷ The turbine changes thermal energy from the hot steam into kinetic energy.

❸ The generator changes kinetic energy into electrical energy.

Power Station Fuels

Fossil fuels...
- are formed from dead plants and animals over many, many years
- will eventually run out, maybe during your lifetime, which is why they are known as non-renewable energy sources
- release polluting gases into the atmosphere when they're burned.

Biofuel (biomass)...
- will not run out because more plants can be grown to make more fuel, so biofuel is known as a renewable energy source
- releases less polluting gases into the atmosphere overall.

Scientists and engineers are currently looking for new energy sources that are renewable and will not pollute the environment.

Other Methods of Producing Energy

Renewable energy sources can be used, but they don't produce very much energy:
- Wind energy can be used to drive a turbine, which turns a generator. But this relies on the wind blowing. It doesn't pollute the atmosphere, but wind turbines can affect wildlife and cause noise pollution, and some people think they destroy the landscape.
- Solar panels use energy from the Sun to heat water. Solar cells change sunlight into electrical energy.
- Hydro-electricity uses water that runs from rivers or reservoirs to drive turbines, which turn generators in order to produce electrical energy.

Nuclear power stations produce lots of energy, but they also produce radioactive nuclear waste. If there were to be an accident, although unlikely with today's safety standards, it could cause devastation.

Finding Energy

Energy from the Sun

Humans need energy to grow, to move and to keep warm. You get your energy from the food – the plants and animals – you eat.

Plants use energy from the Sun in order to grow. They change light energy from the Sun into useful chemical energy. If an animal eats a plant it gets energy from the plant.

So, whether you eat food that comes from plants, or food that comes from animals, the energy has originally come from the Sun.

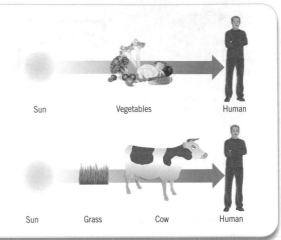

Sun   Vegetables   Human

Sun   Grass   Cow   Human

Energy for Life

Your body changes the chemical energy in the food you eat into the energy you need. Some foods contain more energy than others.

Baked beans – 300kJ   Rice – 600kJ   Portion of chips – 1000kJ

Meeting your Energy Needs

Different people need different amounts of energy, depending on their age, their sex and their lifestyle. The table opposite shows how much energy different people need each day.

People have energy problems if they eat too much or too little food:
- A person who doesn't consume enough energy will lose weight and get tired easily.
- A person who consumes too much energy will gain weight.

It's important to provide your body with the right amount of energy. It's also important to balance your diet with the right amount of vitamins and food types.

Person	Energy Needed Each Day (kJ)
2-year-old child	4000
10-year-old child	8000
14-year-old girl	10000
14-year-old boy	12000
Adult female	10000
Adult male	11000
Adult male with active job	15000

Heat Loss in the Home

Energy is used to heat your home, but some energy will be lost in various ways. On a cold day a badly insulated house can lose up to 10 000J per second. This is wasted energy.

Wasting energy that has been generated in thermal power stations means that more fuel is burned than necessary and more polluting gases are released into the environment.

Polluting gases can contribute to **global warming** and could cause problems for humans, plants and animals.

You can reduce the amount of heat energy lost in a building by improving the **insulation**:
- Double glazing can save about half of the heat lost through windows.
- Cavity walls reduce heat loss. Filling the cavity in the wall with plastic foam insulation reduces heat loss through walls even more.
- Insulating a loft with a 20cm layer of glass fibre can save over half of the energy lost through the roof.
- Draught excluders and foam strips around windows and doors can prevent heat loss.
- Carpets and curtains help reduce heat loss.

Efficiency

Efficiency is a measure of how much energy is used compared with how much energy is wasted in a given situation. For example, a standard light bulb uses 100J electrical energy. It changes 10J into useful light energy, and 90J into wasted heat energy.

You can calculate the efficiency of the light bulb:

$$\text{Efficiency} = \frac{\text{Useful energy}}{\text{Total energy used}}$$

$$= \frac{10J}{100J} = \textbf{10\%}$$

A standard light bulb is 10% efficient. A low-energy light bulb is about 40% efficient, so low-energy light bulbs waste much less energy.

Quick Test

1. What do energy changers do?
2. Give an example of a renewable energy source.
3. What will happen if you consume too much energy?
4. Give one way to reduce energy loss in the home.
5. What formula is used to calculate efficiency?

KEY WORDS
Make sure you understand these words before moving on!
- Chemical energy
- Efficiency
- Energy
- Fossil fuel
- Generator
- Global warming
- Insulation
- Gravitational potential energy
- Joule
- Kinetic energy
- Non-renewable energy
- Renewable energy
- Thermal energy

Finding Energy

Key Words Exercise

Match each key word with its meaning.

Chemical energy	This can't be created or destroyed, only changed
Efficiency	It changes kinetic energy into electrical energy
Energy	This reduces the amount of heat energy wasted
Fossil fuels	Heating of the Earth as a result of polluting gases
Generator	Type of energy resources that can be easily replaced
Global warming	Oil, gas and coal, for example
Insulation	The type of energy possessed by moving objects
Kinetic energy	A way of measuring how much energy is wasted
Renewable energy	The energy stored in food and fuels

Comprehension

Read the passage about James Joule and then answer the following questions.

1. What two types of energy did James Joule first investigate?

2. Give two reasons why other scientists didn't accept James's ideas at first.

3. Why did Michael Faraday and William Thomson eventually accept James Joule's ideas?

4. What is the principle of energy conservation?

James Joule was born near Manchester on 24th December, 1818. As a child, he preferred studying to physical activity. He continued to study in a laboratory he set up at home, even after taking over the family business with his brothers. His investigations led him to work out a relationship between heat energy and electrical energy. He also discovered the principle of energy conservation - that energy can't be created or destroyed, only changed.

Many British scientists would not accept James's ideas, probably because he was only an amateur scientist and his ideas were very different from what most scientists thought at the time. The scientists also found it hard to believe that James could be so accurate with his measurements.

Eventually, James's work on energy was accepted by some famous scientists: Michael Faraday and William Thomson (later known as Lord Kelvin). They could see that James's work fitted in with ideas that other physicists were discovering.

Testing Understanding

1 **Fill in the missing words to complete the sentences about insulating a house.**

a) Most houses are heated by non-renewable fuel such as oil or _____ using

a central heating system, which changes stored _____ energy into

_____ energy. Other houses are heated using electric radiators, which

change _____ energy to _____ energy.

b) A warm house loses heat through the walls, the _____ and the windows.

The heat lost through the _____ can be reduced by using foam insulation

in the _____ between the walls. The heat lost through the

_____ can be reduced by putting glass fibre _____ in

the loft.

c) Heat loss through the windows can be reduced by installing _____

_____. Insulating a house reduces the amount of _____

that needs to be used, saving money and reducing _____ .

2 **Look at the graph about different types of energy then answer the questions that follow.**

The following graph shows the costs of nuclear fuel, a fossil fuel (gas) and wind power for each unit of energy supplied.

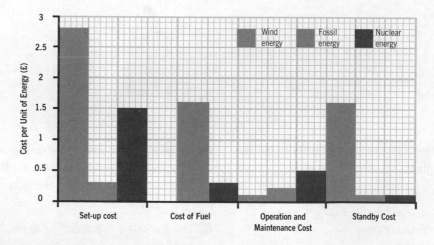

a) Which type of energy has the most expensive fuel?

b) Wind power has no fuel cost. Explain why.

c) The standby cost is the cost of providing energy from somewhere else when the energy supply can't provide energy. Explain why the standby cost for wind power is very high.

d) Neither wind power nor nuclear power creates polluting gases. Why do you think some scientists prefer nuclear power to wind power?

Finding Energy

The diagram shows how you can test a peanut to find its energy content.

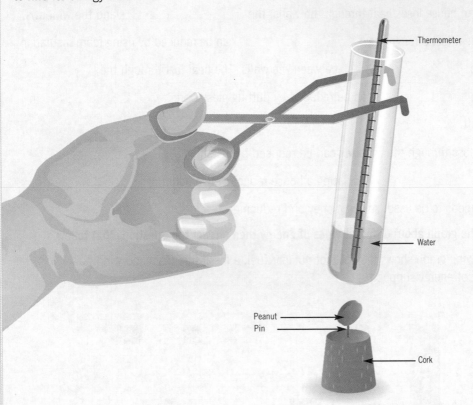

1. Write down instructions for carrying out the experiment.

2. You will also be provided with...
   - a piece of dried bread
   - a marshmallow
   - some butter
   - a small piece of mineral wool to soak the butter in.

   a) How will you compare the chemical energy stored in the different foods?

   b) What will you measure?

3. a) What precautions can you take to try to ensure that your experiment is a fair test?

   b) What are the variables that you must control?

4. List the safety precautions you need to consider.

5. a) What sources of error are there in this method?

   b) How could you reduce the errors?

6. How will you present your results?

Electricity in Circuits

Circuit Symbols

The components in a **circuit** can be represented using symbols. Here are some you should know:

Component	Cell	Battery	Bulb	Switch
Circuit Symbol				
Function	Provides the energy for the circuit in the form of electrical **voltage**.	Two (or more) cells joined together. More cells provide more voltage.	Changes electrical energy in the circuit into light.	Allows the current to be turned on and off.

Simple Circuits

Electricity will only flow in a circuit that is complete. A circuit needs all of the following:

- All the components to be connected in a loop. Circuit 1 will not work because the wire from the bulb back to the cell is missing.
- Connections made of wire to conduct electricity. Circuit 2 will not work because the **conductors** are made of plastic, which doesn't conduct electricity.
- A source of energy, for example, a cell or a battery, to push the **current** around. Circuit 3 will not work because there is no cell.

Circuit 4 is a simple circuit that works. It consists of a cell, a bulb and a switch connected by wires. The circuit diagram is shown below it.

N.B. In the circuit diagram, wires are drawn neatly with straight lines.

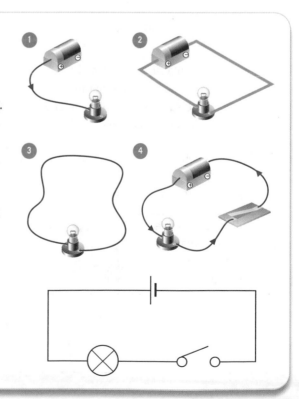

Electricity in Circuits

Current in a Circuit

The electric current in a circuit is a measure of how much electricity is flowing. The current is measured in **amps** (A). It can be measured using an **ammeter**.

In this circuit, a battery of two cells is used to light two bulbs in a single loop. This arrangement is called a **series circuit**. An ammeter has been placed in different positions in this series circuit to measure the current at different places.

By looking at the readings on the ammeter you can see that the current is 0.5A no matter where the ammeter is placed.

In a series circuit, the amount of current flowing is the same at any point in the circuit.

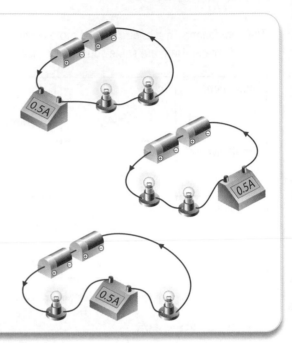

Changing the Current

If more cells are placed in a series circuit, more voltage is supplied. If the number of bulbs is kept the same, the current increases as the number of cells increases:

• With one cell, the current is low and the bulbs are dim.	
• With two cells, the current is higher and the bulbs are brighter.	
• With three cells, the current is highest and the bulbs are very bright.	

If more bulbs are placed in a series circuit, the energy or voltage from the cells has to be shared. If the number of cells is kept the same, the current decreases as the number of bulbs increases:

• With one bulb, the current is high and the bulb is very bright.	
• With two bulbs, the current is lower and the bulbs are dimmer.	
• With three bulbs, the current is very low and the bulbs are very dim.	

Series Circuits

When bulbs are connected in series (in a line) the electricity has to pass through each bulb.

The bulbs glow dimmer in a series circuit because...
- it's more difficult for the electricity to pass through two bulbs than one
- the voltage from the cells (the battery) has to be shared between the two bulbs.

If a switch is included in a series circuit, it will control both bulbs at once:
- If the switch is closed, the circuit is complete and both bulbs will light.
- If the switch is open, the circuit is broken and neither bulb will light.

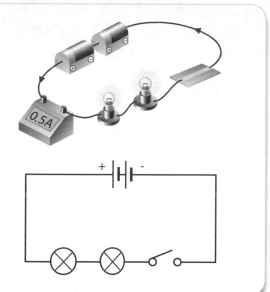

Parallel Circuits

The circuit opposite has two bulbs in it, but this time, the bulbs are connected in parallel.

In a **parallel circuit**, each bulb gets the full battery voltage so the bulbs glow brightly. Together, the two bulbs take more current than in a series circuit so the battery runs down more quickly.

Each bulb has its own loop to the battery. If one bulb is removed there is still a complete circuit from the battery through the other bulb, so it still lights.

Circuit diagram 1 shows how the circuit above would be drawn.

Circuit diagram 2 shows how the circuit would be drawn if two switches were to be added into the circuit so that each bulb can be controlled independently.

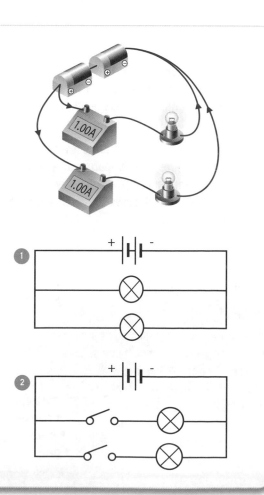

Electricity in Circuits

Safe Electricity

Mains electricity is useful, but it's also very dangerous.

Mains electricity carries much more energy than cells or batteries, so you need to be careful when you use it.

Most accidents are caused by carelessness and can be avoided.

To avoid danger, you should follow these safety precautions when using electricity:

- Don't use **appliances** in the bathroom that are plugged in to the mains.
- Don't turn on a switch or touch a plug when you have wet hands.
- Don't overload a socket with too many appliances; wires may heat up and cause fires.
- Don't use appliances with frayed cables, or try to repair cables with tape – the wires inside will not be insulated properly and can be dangerous.

Fuses

A **fuse** is a thin piece of wire that melts if the current is too high. It's used as a safety device to protect circuits. If the fuse melts, the circuit is broken and the current no longer flows.

Plugs have fuses in them to protect the appliance and prevent overheating. If a fault occurs in an appliance there is usually a surge of high current, which blows the fuse, stopping the flow of current.

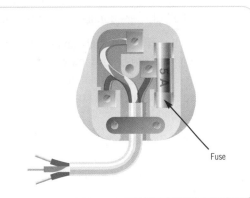

Fuse

Quick Test

1. Draw the circuit symbol for a battery.
2. How can you make a bulb in a series circuit brighter? Give two ways.
3. Is this statement true or false? The battery will run down quicker in a parallel circuit than in a series circuit.
4. What is a fuse?

KEY WORDS
Make sure you understand these words before moving on!
- Ammeter
- Amp
- Appliance
- Battery
- Bulb
- Cell
- Circuit
- Conductor
- Current
- Fuse
- Mains
- Parallel circuit
- Series circuit
- Switch
- Voltage

Key Words Exercise

Match each key word to its meaning.

Ammeter	Allows the circuit to be turned on and off
Amp	Measures the current in a circuit
Appliance	The flow of electricity
Battery	The unit of current
Bulb	A circuit in which everything is connected together in one loop
Cell	Supplies energy for the electricity in the form of voltage
Circuit	A safety device that melts when the current gets too high
Conductor	A material that allows electricity to flow through it
Current	Must be complete for electricity to flow
Fuse	High-energy electricity supplied to your home
Mains	Useful device that uses electricity
Parallel circuit	Changes electrical energy into light and heat energy
Series circuit	The electrical 'push' provided by a cell
Switch	Two or more cells joined together
Voltage	A circuit in which each bulb has its own loop

Comprehension

Read the passage about superconductors and then answer the following questions.

1. What happens to helium at -269°C?

2. What is resistance?

3. What happens to the resistance of some conductors at very low temperatures?

4. Superconductors are used to make very strong magnets. Describe two uses of these strong magnets.

A Dutch physicist called Heike Kamerlinge-Onnes lived in the 1800s. He studied gases at very low temperatures. He cooled helium until it turned into a liquid at -269°C. He then began to cool metals, such as mercury, which conduct electricity.

When a current is passed through a conductor, energy is needed to overcome the conductor's resistance. The resistance opposes the current. Heike found that when metals are cooled to very low temperatures their resistance reduces to almost zero.

These materials are called superconductors. Superconductors are used to make strong magnetic fields in MRI scanners in hospitals and in Maglev trains (high-speed trains suspended above the track by magnets).

Electricity in Circuits

Testing Understanding

1 **Fill in the missing words to complete the sentences about electrical circuits.**

a) Everything is connected in one single loop in a _____ circuit. If a switch is placed in this type of circuit with two bulbs, the switch will control _____ bulbs.

b) If one of the _____ blows in a series circuit, the other bulb will _____ because the blown bulb will have _____ the circuit.

c) In a _____ circuit each bulb has its own loop to the _____. Two _____ can be placed in a parallel circuit to control the bulbs independently.

d) If one of the bulbs blows in a parallel circuit, the other bulb _____ because it has its own loop to the battery. In parallel circuits the _____ are brighter, but the battery will run down _____.

2 **Draw a circuit diagram for each of the following circuits.**

The symbol for an ammeter is —(A)—.

a)

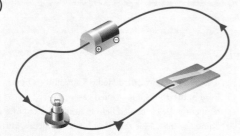

b)

c)

d)

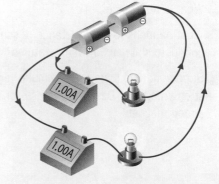

Sara decides to investigate what happens to the current in a parallel circuit when the number of bulbs in the circuit is increased.

She has two batteries, three bulbs, three switches and four ammeters.

She sets up three circuits:
- Circuit A has a battery of two cells, one bulb and a switch.
- Circuit B is a parallel circuit. It has a battery of two cells, two bulbs and two switches.
- Circuit C is a parallel circuit. It has a battery of two cells, three bulbs and three switches.

She records the current through each circuit. Here are her results:

Circuit	Current through each bulb	Total current
A	0.25A	0.25A
B	0.25A	0.50A
C	0.25A	0.75A

1 a) Draw a circuit diagram for a Circuit A.
   b) Add an ammeter symbol to your circuit diagram to show where the ammeter could be connected so the current through the bulb can be measured.

2 a) Draw a circuit diagram for a Circuit C.
   b) Add four ammeter symbols to your circuit diagram to show where the ammeters could be connected so the current through each of the three bulbs and the total current can be measured.

3 Look at Sara's results table. What happens to the current in a parallel circuit? What effect does increasing the number of bulbs in a parallel circuit have on the current in a circuit?

4 In which circuit, A, B or C do you think the cells will run out the fastest? Give a reason for your answer.

5 Why is it important for Sara to disconnect the circuit as soon as she has taken her readings?

What Forces Do

Forces

There are many different types of **forces** and they can do different things.

Although you can't see a force, you can see its effect:

- A force can make an object move, e.g. the **tension** (strain force) of a string on a bow makes an arrow move.
- A force can make an object stop moving, e.g. the **impact force** of a goalkeeper's hands on a ball stops it moving.
- A force can change the speed of a moving object, e.g. the force of a car engine's **thrust** makes it go faster.
- A force can change the direction of a moving object, e.g. the impact force of a racket changes the direction of a tennis ball.
- A force can change the shape of an object, e.g. you can change the shape of an orange by squeezing it.

Measuring Forces

Forces are measured in units called **newtons**.

You can use a newton meter to measure a force: the further the spring stretches, the larger the force. A scale and a pointer show the size of the force.

- A force of about 1N is needed to lift an apple.
- A force of about 15N is needed to pull open a door.
- A force (thrust) of about 800 000N might be exerted by a jet engine in a plane.

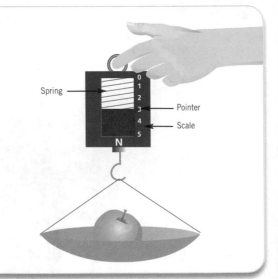

Spring — Pointer — Scale

Contact Forces

Forces that act when two objects touch are known as **contact forces**.

Impact force, strain force and friction are examples of contact forces.

Impact Force

An impact force is experienced when two objects bump into each other, for example...
- when a particle in the air hits your skin
- when two moving cars crash into each other
- when a tennis ball hits a racket.

Strain Force

A strain force (tension) is experienced when an object is stretched, for example...
- when an elastic band or a spring is stretched
- when a picture hangs on a wall by a string
- when a ball hits the strings of a tennis racket.

Friction

Friction is the force that slows down a moving object:
- Smooth surfaces reduce the friction between objects.
- Rough surfaces increase the friction between objects.

Friction can be a useful force, e.g. friction between your foot and the ground allows you to walk; if there were no friction you would slip.

In some cases, it helps if friction is increased, for example...
- footballers wear studs on their trainers to increase the friction between their feet and the field
- larger brake pads on a bicycle wheel create more friction, slowing it down more quickly.

In other cases, it's better if friction is reduced. For example, skiers wax the bottom of their skis to reduce friction so that they can go faster.

Friction generates heat, for example...
- when you rub your hands together they get warm
- when the moving parts in a car engine rub together the car can get too hot and overheat.

Friction causes particles to heat up when they travel at high speed through the upper atmosphere:
- Friction causes dust particles in comets' tails to light up, producing shooting stars.
- Friction can be a problem when shuttles and satellites are sent into space because it causes them to get very, very hot.

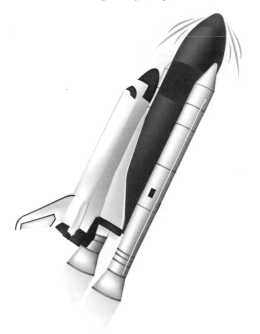

What Forces Do

Non-Contact Forces

Non-contact forces act without touching an object. The region in which a non-contact force acts is called a **field**.

Magnetic force, electrostatic force and gravitational force are examples of non-contact forces.

Magnetic Force

Some materials experience a magnetic force when a magnet is near them.

A magnet can exert a pulling force on some materials without touching it. The closer the magnet is, the stronger the force is.

Two magnets can also exert a pushing force on each other.

Electrostatic Force

An electrostatic force can be created when some materials are rubbed together. For example, if you rub a comb on a jumper you will create static electricity.

Two charged objects exert a force on each other without touching each other. They can **either**...
- push each other apart **or**
- pull each other together.

For example, if you hold a charged comb near water running from a tap, you can bend the flow of water using the electrostatic force.

Gravitational Force and Weight

Gravitational force is the force of attraction between any two masses in the Universe, eg...
- the force between your body and a book is very small
- the gravitational force between planets is large.

On Earth all masses are attracted to the Earth itself. The size of this force depends on the **mass** of the object.

The unit of mass is the kilogram.

The gravitational force acting on any mass is called **weight** and is measured in newtons, like all forces.

The weight of an object can be calculated using the following equation:

Weight = Mass x 10N/kg

> 10N/kg is the gravitational field strength on Earth.

Example
Calculate the weight of a sports bag that has a mass of 1.5kg.

Weight = 1.5kg x 10N/kg = 15N

N.B. It is wrong to say 'My weight is 65kg'. You should say 'My mass is 65kg and my weight is 650N'.

Speeding Up, Slowing Down

When a force is applied to a **stationary** object it may begin to move. For example, when a footballer kicks a ball, it moves.

When a force is applied to a moving object in the opposite direction of its motion, it will slow down or even stop. For example, a football may be stopped…

- quickly by the impact force of the goalkeeper
- slowly by the contact force of friction as it rolls on the ground.

When a force is applied to a moving object in the same direction as its motion, it will speed up.

Balanced Forces

A skydiver experiences two forces: the gravitational force of weight downwards and the force of friction with the air (**drag**) upwards.

If these forces are of equal size then the skydiver neither speeds up nor slows down. She will continue falling at the same speed. The forces are said to be **balanced**.

Upthrust

When an object is in a liquid it experiences an upward force from the water called **upthrust**.

The upthrust is caused by the water being pushed out of the way. Upthrust is determined by the volume of the object and the density of the liquid.

The upthrust makes things weigh less when they're in a liquid. If the upthrust is large enough to balance the weight of the object, then the object will float.

Density

Density is a measure of how heavy something is for its size. It can be calculated using this equation:

$$\text{Density} = \frac{\text{Mass}}{\text{Volume}}$$

Example
Calculate the density of a cube of wax that has a volume of 2cm^3 and a mass of 1.9g.

$$\text{Density} = \frac{1.9\text{g}}{2\text{cm}^3} = 0.95\text{g/cm}^3$$

You can use density to predict whether an object will float or sink when it is placed in a liquid. For example, the cube of wax will float in water because it is less dense than water. (Water has a density of 1g/cm^3.)

Quick Test

1. What unit is used to measure force?
2. Two cars crash into each other. What type of force is this an example of?
3. A skydiver jumps out of a plane. What happens to the skydiver's speed when the forces acting on him are equal?
4. What is density a measure of?

KEY WORDS
Make sure you understand these words before moving on!
- Balanced forces
- Contact force
- Drag
- Field
- Friction
- Impact
- Mass
- Newton
- Stationary
- Tension
- Thrust
- Upthrust
- Weight

What Forces Do

Key Words Exercise

Match each key word to its meaning.

Balanced forces	The force when a spring or a rubber band is stretched
Contact force	The area where a non-contact force acts
Drag	1kg of this has a weight of 10N
Field	Forces that involve objects touching
Impact	A contact force when objects collide
Mass	Not moving
Stationary	A force provided by an engine
Tension	The upward force from a liquid that can cause an object to float
Thrust	When the forces acting on an object cancel out each other
Upthrust	The pull of the Earth on a mass due to gravity
Weight	Friction between a moving object and the air

Comprehension

Read the passage about the motion of objects and then answer the following questions.

1. What did the ancient Greeks believe was pushing an object such as an arrow?

2. What did Newton say is needed to make a stationary object move?

3. What did Newton say is needed to make a moving object stop?

4. What would happen to a moving arrow if no force acts on it?

5. Why do many people still believe that objects stop on their own, without a force?

Ancient Greeks believed that objects, for example, an arrow moving horizontally, were pushed by the air, and the object fell to the ground when the air became tired of pushing.

Sir Isaac Newton realised that things didn't keep moving until the air stopped pushing it. He published a book in 1686 that described the relationships between the forces acting on a body. One of his laws of motion says the following:

- If something is stationary, you have to apply a force to make it move.
- If something is moving it tends to keep moving at a steady speed in a straight line forever unless another force is applied.

Even today many people are confused by this idea; people often think that things need to be pushed in order to keep moving. On Earth, moving objects are usually stopped by the force of friction. So it seems strange to say that something moving will keep moving forever unless a force stops it. In space, though, where there is no friction (not even air friction because there is no air) moving objects can keep moving forever!

1 **Fill in the missing words to complete the sentences about forces.**

   a) The force that acts against all moving objects is called _____. This is a

        _____ force because the objects are touching.

   b) When a force acts on an object without touching it, the force is called a _____

        force. The area where this kind of force acts is called a _____. Examples

        of these forces are _____ force, electrostatic force and

        _____ force.

   c) All forces are measured in _____. The _____ of an

        object depends on its mass and the gravitational field strength.

   d) An object of mass 30kg would have a _____ on Earth of _____.

2 **Read the information then answer the questions that follow.**

An experiment was carried out in which
the length of a spring was measured when
different weights were applied to it.

The table below shows the results of
the experiment.

Force	Length of Spring
2	5.0
3	6.5
4	8.0
5	9.5
6	11.0
7	12.5

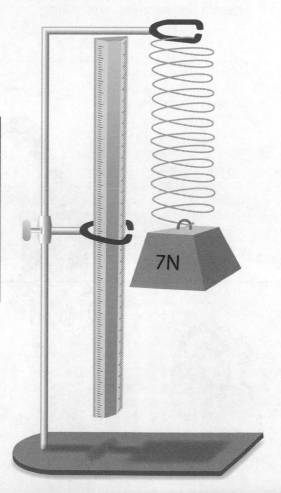

   a) **i)** What unit would force be measured in?
        ii) What unit would length be
           measured in?
   b) Plot a graph of force against length for
        the spring.
   c) Describe the relationship between the
        force and the length of the spring.
   d) From the graph, find the length of the
        spring when no force is applied.

What Forces Do

Skills Practice

David has been asked to carry out an investigation to measure the density of plasticine.

He has been provided with...
- some plasticine
- some fishing line
- a measuring cylinder
- some water
- a pan balance.

1 David will use the fishing line to lower the plasticine into the measuring cylinder. How will this improve his experiment?

2 Draw a diagram to show how David will measure the volume of the plasticine using the water and the measuring cylinder.

3 List three precautions David should take to reduce error in his method.

4 David carries out the experiment three times and finds that the plasticine has an average volume of $2.0cm^3$. He uses the pan balance and finds that the plasticine has a mass of 3.4g. Find the density of the plasticine

5 Would you expect the density of water to be greater than or less than the density of plasticine? Explain your answer.

 For further practice, go to p.36-39 of the Year 7 Science Essentials Workbook

Beyond Our Planet

The Earth

The Earth is a relatively small, cool ball called a **planet**.

The Earth spins around slowly on its **axis**. It completes one full rotation every 24 hours. This is one day.

As the Earth turns, different places on it move from the sunlight into the shadow. This gives us day and night. The Earth moves around the **Sun** in a path called an orbit once every $365\frac{1}{4}$ days. This is one year.

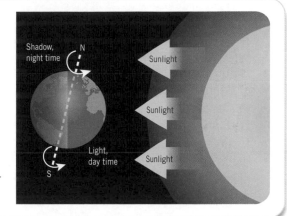

Seasons

The Earth rotates on a tilted axis.

This means that at different times in its orbit, different **hemispheres** are tilted towards or away from the Sun.

This changes the area that the sunlight shines on.

- When the Southern hemisphere is tilted towards the Sun it is summer in the Southern hemisphere and winter in the Northern hemisphere.
- When the Northern hemisphere is tilted towards the Sun it is summer in the Northern hemisphere and winter in the Southern Hemisphere.

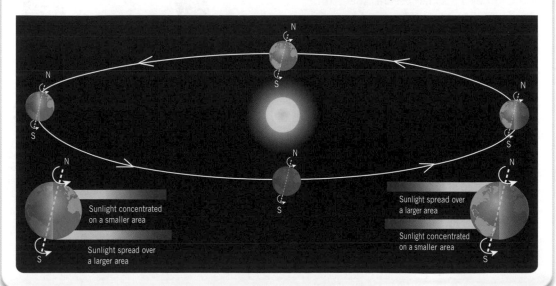

Beyond Our Planet

The Moon

The **Moon**...
- is smaller than the Earth
- is covered with rocks and craters
- is 380 000km away from the Earth
- orbits the Earth approximately every 28 days
- rotates on its axis – its 'day' is 28 days (This means that the same side of the Moon is always facing the Earth).
- doesn't give off its own light – we see it at night because its surface reflects the sunlight shining on it.

At different points in its orbit, the amount of the Moon you can see changes. This is known as the Moon's **phases**.

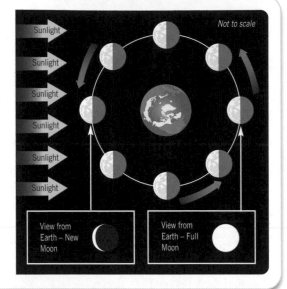

Satellites

There are hundreds of **satellites** orbiting the Earth. Many of these satellites are in **geostationary orbits**. This means that they orbit the Earth at the same speed as the Earth rotates, so they always stay above the same point on the Earth.

Satellites are used for many different things:
- Communication satellites receive and send television and telephone signals.
- Satellites are used to take pictures of weather patterns so that forecasters can predict the weather.
- Navigation satellites send out signals about position to ships, aircraft and satellite navigation systems in cars.
- Research satellites carry telescopes and cameras to send pictures of the Universe back to Earth.

Our Solar System

When ancient civilisations studied the movement of the Sun, Moon, **stars** and planets across the sky they believed that all these objects moved in circular paths around the Earth.

Copernicus (1473–1543), from Poland, suggested that the motion of the planets could be explained better if they were all rotating around the Sun.

The Sun doesn't look very big from Earth, because it's 150million kilometres away from us!

In our **Solar System** there are nine planets that orbit the Sun at different distances. This table gives some information about the planets.

Planet	Diameter (km)	Approximate Distance from Sun (million km)
Mercury	4 880	58
Venus	12 100	108
Earth	12 800	150
Mars	6 800	230
Jupiter	143 000	780
Saturn	120 000	1430
Uranus	51 000	2870
Neptune	48 600	4500
Pluto	5 500	5910

N.B. Pluto is now known as a dwarf planet

In between Mars and Jupiter there is a band of rocks and debris called the **Asteroid Belt**.

The picture opposite is not to scale.

To get some idea of the scale, imagine a football at one end of a football pitch and a pea at the other end. This approximately represents the relative sizes of the Earth and the Sun and the distance between them.

Beyond Our Planet

Galaxies

The Universe is made up of millions of **galaxies**. Galaxies are spread out with big spaces between them, and they are all moving away from each other.

Our galaxy is known as the Milky Way. It is a spiral galaxy. The distance from one side of the Milky Way to the other is about 100 000 light years. (1 light year is the distance a beam of light can travel in a year. It is equal to about 9.5 million, million kilometres.)

There are millions of stars in each galaxy. The Sun is about two thirds of the way out from the centre of our galaxy. The nearest star to our Sun is Proxima Centauri which is 4.2 light years away.

Stars are huge glowing balls of gas that give out heat and light. During the daytime the Sun is so bright you can't see the other stars.

About 1500 stars in the Milky Way are bright enough to be visible after the Sun has set.

Many of the stars are as bright as our Sun, but they are much further away, which makes them appear much smaller and dimmer.

Constellations

Stars seem to make patterns and shapes in the sky. These patterns are called **constellations**.

The Plough (shown opposite) is one such constellation. If you follow the two end stars of the plough they point to the Pole Star.

Whilst most stars appear to move across the night sky as the Earth turns, the Pole Star stays still because it is exactly in line with the North Pole and the axis on which the Earth turns.

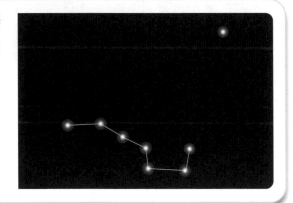

Quick Test

1. How long does it take the Earth to complete one rotation on its axis?
2. How long does it take the Earth to complete one orbit around the Sun?
3. Give two uses of satellites.
4. Which planet is closest to the Sun?

KEY WORDS
Make sure you understand these words before moving on!
- Asteroid Belt
- Axis
- Constellation
- Galaxy
- Geostationary orbit
- Hemisphere
- Moon
- Phase
- Planet
- Satellite
- Solar System
- Star
- Sun

Key Words Exercise

Match each key word to its meaning.

Asteroid belt	An object that orbits a planet
Axis	A pattern of stars that can be seen in the night sky
Constellation	The visible part of the Moon at night that changes during a month
Galaxy	An orbit in which the satellite stays in a fixed position relative to the Earth
Geostationary orbit	A natural satellite that orbits the Earth once every 28 days
Hemisphere	The Milky Way, for example
Moon	A fiery ball of burning gas found at the centre of our Solar System
Phase	Object that orbits a star; the Earth, for example
Planet	Area between Mars and Jupiter, where lumps of rock are found orbiting the Sun.
Satellite	The collection of our Sun and the planets
Solar System	Balls of burning gas found in clusters called galaxies
Stars	Half of the Earth, either the North or the South
Sun	The line around which a planet rotates

Comprehension

Read this passage about the life of a star and then answer the following questions.

1. What is a luminous object?

2. What force pulls together the dust and gases that form a star?

3. What causes the high temperatures in a star?

4. What is a supernova?

5. What size of star is likely to become a black hole?

Stars are luminous objects; they give out their own light. They are formed from giant clouds of dust and gas drawn together by the force of gravity. Our Sun is an average star. The temperature at its centre is about 15 000 000°C and at the surface it is about 6000°C. These high temperatures are caused by nuclear reactions in the star.

Stars don't last forever. They change very slowly over millions of years. Some stars die quietly, gradually cooling until they don't emit any light. Larger stars die with an enormous explosion called a supernova. A supernova can be 500 million times brighter than the Sun. Stars that are even larger may collapse in on themselves and become a black hole, from which nothing can escape.

Beyond Our Planet

Testing Understanding

1 Fill in the missing words to complete the sentences about orbiting bodies.

a) The Earth takes 24 _____ to rotate on its axis. This is known as one

_____.

b) The Earth takes 365 $\frac{1}{4}$ days to _____ the Sun. This is known as a

_____.

c) The Earth rotates on a tilted axis.

   i) For half of the year the Northern _____ is tilted towards the Sun.

      During this time, the season in the Northern hemisphere is _____,

      and in the Southern hemisphere it is _____.

   ii) For the other half of the year the _____ hemisphere of the Earth is

      tilted towards the Sun. During this time, the season in the Southern hemisphere is

      _____, and in the Northern hemisphere it is _____.

d) The _____ orbits the Earth approximately every 28 days. It rotates on

its _____ every 28 days; this means that the same side always faces

the _____.

2 Look at the graph, then answer the following questions.

The graph shows the approximate time it takes for some of the planets in the Solar System to rotate on their axis.

a) Which planet is the Earth? Give a reason for your answer.

b) Saturn and Jupiter take the same amount of time to rotate. Which planets are Saturn and Jupiter?

c) Mars rotates more slowly than the Earth. Which planet is Mars?

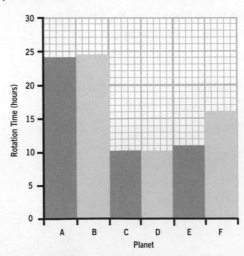

Skills Practice

You have been asked to investigate the relationship between the average surface temperature of the planets and their distance from the Sun. You have been given the following data.

Planet	Distance from Sun (million km)	Time taken to Complete One Orbit	Diameter (km)	Average Surface temperature (°C)
Earth	150	1 year	12 800	22
Mars	230	1.9 years	6 800	-23
Jupiter	780	12 years	143 000	-140
Saturn	1430	29 years	120 000	-180
Uranus	2870	84 years	51 000	-210

1 What variables would you plot on a graph to test this relationship?

2 a) Copy the axes opposite and plot a graph to test this relationship. Remember to add labels to the axes.
b) Carefully draw a smooth curve through your points.

3 How does the average surface temperature of a planet change with its distance from the Sun?

4 a) Venus is 110 000 000km from the Sun. What average surface temperature would you expect Venus to have?
b) The average surface temperature of Venus is actually 480°C. This does not fit very well on the curve. What other factors might affect the average surface temperature of a planet?

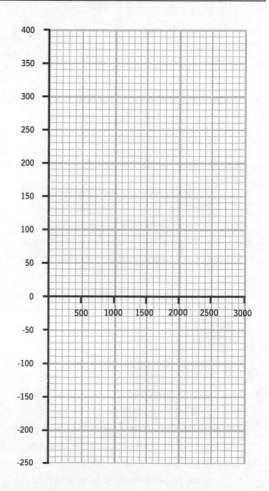

Getting Hotter, Getting Colder

Temperature

Humans aren't good judges of **temperature**. The boy opposite has his left hand in hot water and his right hand in cold water. When he puts both hands into the bowl of warm water, his left hand feels cold and his right hand feels warm.

The hotness or coldness of something is measured by taking its temperature.

The unit of temperature is **degrees Celsius** or °C.

You can measure temperature accurately using a **thermometer**:

- Galileo invented the first thermometer in 1600 based on the expansion of air.
- In the 1700s, Daniel Fahrenheit, a German scientist, invented a better thermometer using expanding mercury in a glass tube. His unit for temperature was the degree Fahrenheit.
- The Celsius scale of temperature is named after Anders Celsius, a Swedish scientist. His scale placed melting ice at 0 and boiling water at 100, which made it easier for people to make a thermometer.

COLD WATER WARM WATER HOT WATER

Heat

Temperature isn't the same as **heat**. Heat...

- is a form of energy. A full bath at 25°C has more heat energy than a cup of tea at a higher temperature of 75°C. This is because the bath holds a greater volume of water
- flows from hot to cold. If there's a difference in temperature between an object and its surroundings, there's a flow of heat from hot to cold.

When the boy above puts his hands into the warm water, he feels cold in his left hand because heat is flowing away from his hot hand into the warm water. His right hand feels warm because heat is flowing from the warm water into his cool hand.

Conduction

Heat travels through solids by a process called **conduction**. If you stir hot soup with a metal spoon, the heat quickly flows through the metal from the soup to your hand. Metal is a good conductor because heat can flow easily through it. As the particles in the solid vibrate, the heat energy is passed along.

If you stir the hot soup with a wooden or a plastic spoon, the heat can't travel so easily to your hand. Wood and plastic aren't good conductors of heat; they're called **insulators**.

A stone floor will feel colder to your bare feet than a carpet because...

- stone is a good conductor of heat and the heat flows quickly from your feet into the stone
- a carpet is an insulator and doesn't conduct the heat away from your feet quickly.

Convection

Heat can travel through liquids and gases:

- In a kettle, only the water at the bottom of the kettle is heated but the heat travels through all the water until it boils.
- When a room is heated using a radiator, only the air next to the radiator is heated but the heat travels until all of the room is heated.

Liquids and gases are poor conductors of heat. Heat travels through them another way, by **convection**.

Convection...

- is when the particles in a liquid or a gas move, carrying the heat with them from one place to another
- can't happen in solid objects because the particles aren't free to move from one place to another.

The movement of the air around a room is called a convection current. Hot air carries the heat and the cooler air travels back towards the heat source.

Getting Hotter, Getting Colder

More About Convection

Hot air expands, making it less dense and therefore lighter than cold air. Hot liquids are also lighter than cold liquids. This means that hot air or liquid rises:

- Hot-air balloons rise.
- In a kettle, the water at the bottom is heated. It becomes lighter and moves upwards. The cold water takes its place next to the element and heats up.
- In a fridge, the cooling element is placed at the top. The air is cooled by the element and sinks to the bottom of the fridge because it's heavier. The hotter air takes its place next to the element and is cooled. The air cycles until the fridge is cool.

Convection currents also occur in nature:

- Land heats up from the Sun and heats the air above it. This air rises in currents called thermals.
- Gliders and birds use thermals to gain height.

Keeping Warm

Most houses waste energy. When the temperature is cold outside, energy is lost by heat escaping, mostly by convection.

The wasted heat can be reduced by insulating the house:

- Double glazing traps air between layers of glass.
- Curtains trap air.
- Loft insulation reduces the amount of hot air escaping through the roof.
- Cavity walls trap air. Cavities are often filled with insulation that reduces air movement.
- Draught proofing.
- Reflective foil on or in walls reduces heat loss by radiation.
- Insulation (lagging) around the hot-water tank.

In hot weather, a well-insulated house will also stay cooler by reducing the amount of heat coming into the house from outside.

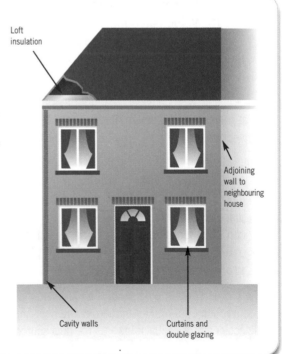

Loft insulation

Adjoining wall to neighbouring house

Cavity walls

Curtains and double glazing

Radiation

Heat can travel when there are no particles to carry it. **Infrared** waves carry heat energy through air or through a vacuum. This is called radiation.

If you put your hand near something hot, you can feel the heat on your skin without touching it. The hotter the object, the more infrared energy it radiates. This is how the Sun's energy reaches the Earth.

Some infrared radiation can pass through glass:
- Infrared radiation from the Sun can pass through glass into a greenhouse because the Sun is very hot and it radiates high-energy radiation.
- The plants inside the greenhouse aren't so hot and radiate infrared radiation of lower energy that can't pass through the glass.
- The energy is trapped and the greenhouse heats up.

The Earth's atmosphere acts like the glass in a greenhouse, trapping energy from the Sun. Without this greenhouse effect, the Earth wouldn't be warm enough for life to exist.

However, polluting gases such as carbon dioxide (for example, from power stations) can increase the effect. This extra warming is known as **global warming** and could cause problems for humans, plants and animals.

Darker colours radiate more energy than lighter colours. Dark, dull surfaces also absorb more infrared energy than light, shiny surfaces.

Shiny surfaces reflect radiation. This is why on a sunny day it's a good idea to wear light colours to keep cool.

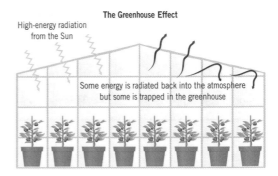

The Greenhouse Effect

High-energy radiation from the Sun

Some energy is radiated back into the atmosphere but some is trapped in the greenhouse

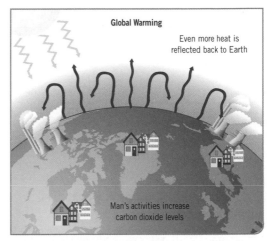

Global Warming

Even more heat is reflected back to Earth

Man's activities increase carbon dioxide levels

Quick Test

1. What unit is temperature measured in?
2. Which has more heat: a full bath at 25°C or a cup of tea at a temperature of 75°C?
3. In which direction does heat flow?
4. Name three ways in which heat travels.
5. What kind of waves carry heat energy?
6. Why is it better to wear light colours on a hot day?

KEY WORDS

Make sure you understand these words before moving on!
- Conduction
- Convection
- Degrees Celsius
- Global warming
- Heat
- Infrared
- Insulation
- Radiation
- Temperature
- Thermometer

Getting Hotter, Getting Colder

Key Words Exercise

Match each key word with its meaning.

Conduction	The way heat flows through a solid
Convection	A type of energy
Degrees Celsius	Heat radiation waves
Global warming	This reduces the amount of heat flow
Heat	An instrument used to measure temperature
Infrared	The Earth getting hotter as a result of the greenhouse effect
Insulation	The unit of temperature
Radiation	A measure of the hotness of a body
Temperature	The way heat travels through air or a vacuum
Thermometer	The way heat travels through gases or liquids

Comprehension

Read the passage about getting cold, then answer the following questions.

1. Where does the human body get the energy from to keep warm?

2. What percentage of energy is converted to heat when a muscle is moved?

3. At what temperature does the body stop working properly?

4. Describe two ways that heat loss can be reduced in a person suffering from hypothermia.

5. Describe two ways that the human body reacts when a person is suffering from hypothermia.

The human body keeps warm by transferring chemical energy in food to heat energy in the muscles. When a muscle moves, only about 15% of the energy used actually moves the muscle; the rest is converted to heat.

Mountain climbers keep warm because they're doing lots of exercise. Their hands and feet may get very cold but they can cope with the conditions, provided their body's core temperature doesn't drop below 37ºC. If their core temperature drops below 35ºC, their body stops working properly. This condition is known as hypothermia.

Lots of layers of clothing trap air and prevent heat loss by convection. A shiny blanket can be wrapped around someone with hypothermia to reflect the heat inwards and prevent heat loss by radiation.

A cold body will shiver to generate more heat in the muscles and the blood vessels will shrink so that heat isn't carried to the skin but is kept near important organs, such as the heart.

Testing Understanding

1 **Fill in the missing words to complete the sentences about heat on the move.**

a) Conduction is the method of heat flow in _____. When you stir soup

with a metal spoon, heat is _____ through the spoon and into your hand.

Metal is a _____ conductor.

b) If you stir the soup with a plastic or_____ spoon, the heat isn't transferred

through the spoon to your hand so quickly. These materials are known as _____.

c) In _____ and gases, heat travels by _____. Heat can't

travel by convection in a solid. In air and in a _____, heat can travel by

_____.

2 **Read the information below about the Thermos flask, then answer the questions that follow.**

The Thermos flask was invented by James Dewar about 100 years ago. It's usually used to keep drinks hot. Look at the diagram opposite of his first flask.

a) The vacuum prevents heat travelling in two ways. What are they?

b) What does the shiny silver surface do?

c) Fill in the missing words to complete the following sentence:

The stopper prevents evaporation and

hot air rising out of the flask. Hot air rising

is known as a _____

_____.

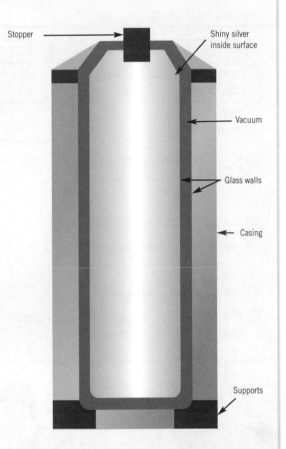

Stopper

Shiny silver inside surface

Vacuum

Glass walls

Casing

Supports

d) Modern flasks are very similar, but are often made of metal instead of glass because metal is stronger. What is the main disadvantage of using metal instead of glass?

e) Would a Thermos flask keep a cool drink cool? Explain your answer.

Getting Hotter, Getting Colder

Skills Practice

Richard and Sarah are investigating the cooling of hot water in beakers with different insulation.

Sarah insulates her beaker with layers of cotton wool. Richard insulates his beaker with layers of aluminium foil. They stir the water as it cools and take the temperature every 30 seconds.

Here are their results:

Time	0	30	60	90	120	150	180	210	240	270	300
Temperature (cotton wool)	80.0	76.0	73.5	71.0	69.0	67.5	66.0	65.0	64.5	64.0	63.5
Temperature (aluminium foil)	80.0	74.0	71.0	68.0	66.0	64.0	62.5	61.0	60.5	60.0	59.5

1. Name three things that must be kept the same for both beakers in order for the experiment to be a fair comparison.

2. Explain how cotton wool insulates the beaker.

3. Explain how aluminium foil insulates the beaker.

4. Add units to the table above.

5. Plot both sets of results on the same axes. Draw a smooth curve through each set of results.

6. Which material provides the best insulation?

7. Why do you think that some of the points don't lie exactly on the smooth curve? What could be done during the experiment to try to ensure that the line is smoother?

8. Sanjay did the same experiment, but his layers of aluminium foil weren't wrapped so tightly. There were air gaps in between the layers. Do you think this would make the foil a better or a poorer insulator? Explain your answer.

For further practice, go to p.12-15 of the Year 8 Science Essentials Workbook

Magnetism and Electromagnetism

These materials have magnetic properties, so they can be magnetised:

- Iron (a metal).
- Cobalt (a metal).
- Nickel (a metal).
- Steel (an alloy).

Even if these materials aren't magnetised, they're still **attracted** by other magnets.

Non-magnetic materials, such as plastic, wood and other metals, can't be magnetised and aren't attracted by magnets.

Magnetic Fields

The strongest forces from a magnet seem to come from the ends, the north-seeking pole (or north pole) and the south-seeking pole (or south pole) of the magnet.

The space around a magnet is called a **magnetic field**. You can study the field using iron filings:

- The magnetic **field lines** run from the north pole to the south pole.
- The magnetic field is strongest where the field lines are closest together.

The Earth has a magnetic field and pulls on the poles of a magnet. A compass is a tiny magnet that turns on a spindle. The north-seeking pole turns so that it points to the north pole. A bar magnet suspended from a string does the same.

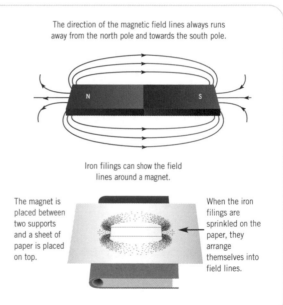

The direction of the magnetic field lines always runs away from the north pole and towards the south pole.

Iron filings can show the field lines around a magnet.

The magnet is placed between two supports and a sheet of paper is placed on top.

When the iron filings are sprinkled on the paper, they arrange themselves into field lines.

Attracting and Repelling

A magnet will always attract another magnetic material (e.g. an iron or steel bar) that isn't magnetised.

Two magnets will either attract or **repel** one another, depending on how they're arranged. Unlike poles attract but like poles repel.

Two like magnetic poles brought near to each other will repel each other.

Two unlike magnetic poles brought near to each other will attract each other.

Magnetism and Electromagnetism

Making a Magnet

A magnetic material...
- is full of tiny magnetic areas called **domains**.
- becomes magnetised when all the domains are lined up in the same direction.

Once a magnet is magnetised, it generally stays magnetised unless its magnetism is destroyed. Soft magnetic materials don't retain their magnetism because their particles rotate back to their random positions.

If a magnetic material is held near another magnet, it may become magnetised. Stroking a magnetic material repeatedly in the same direction with a magnet will also magnetise it.

Each paper clip in the chain opposite has been magnetised and attracts the next paper clip.

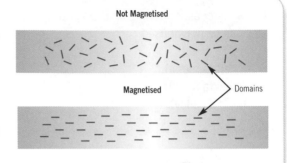

Not Magnetised

Magnetised

Domains

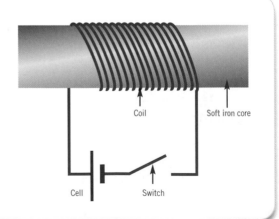

Electromagnets

A **coil** of wire will become magnetic when a direct current is passed through the coil. This is called an **electromagnet**.

Adding a **core** makes the electromagnet stronger. If the core is made of a soft magnetic material, such as iron, the magnet will 'switch off' when the current is turned off.

An electromagnet can also be made stronger by...
- increasing the number of turns of the coil
- increasing the current.

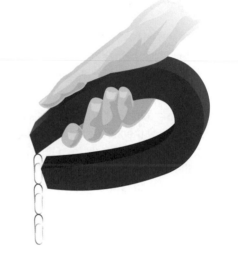

Coil

Soft iron core

Cell

Switch

Destroying a Magnet

A magnet becomes demagnetised when all the domains return to random positions. This can be done by...
- banging it with a hammer or dropping it
- heating it
- passing an alternating current through a coil of wire wrapped around the magnet.

Breaking a magnet in two doesn't destroy it; it simply creates two smaller magnets.

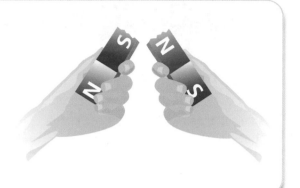

Uses of Magnets and Electromagnets

An electromagnet can do all the things that an ordinary magnet can, but it can also be switched on and off. The electromagnet opposite is being used to move a car in a scrap yard.

Electromagnets and magnets are used in many everyday devices including **circuit breakers**, electric bells, electric motors, loudspeakers, generators in power stations, credit cards and imaging scanners in hospitals.

A Maglev train is suspended above the rails by strong magnets, reducing friction.

The Circuit Breaker

A circuit breaker is a safety device that cuts off the current in a circuit if it gets too high:
- The current flows through the contact switch and the electromagnet.
- If the current gets too high, the pull of the electromagnet is strong enough to pull the soft iron bar to the right.
- The movement of the soft iron bar allows the reset button, attached to the contact switch, to spring upwards, breaking the circuit and stopping the current.
- Once the problem in the circuit has been fixed, pushing the reset button allows the current to flow again.

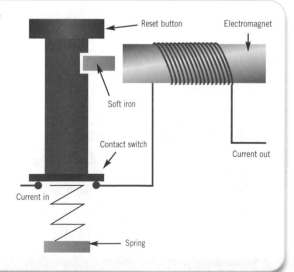

Reset button

Electromagnet

Soft iron

Contact switch

Current out

Current in

Spring

Magnetism and Electromagnetism

The Electric Bell

An electric bell works along the following principle:

- When the switch is closed, the electromagnet is switched on and attracts the soft iron.
- The movement of the soft iron pulls the hammer, which strikes the bell.
- When the hammer moves, the contacts are pulled apart, breaking the circuit.
- The current stops flowing and the electromagnet loses its magnetism.
- The contacts are pulled back together by the spring, switching the electromagnet back on.
- The repeated cycle of the hammer striking the bell causes the ringing of the bell.

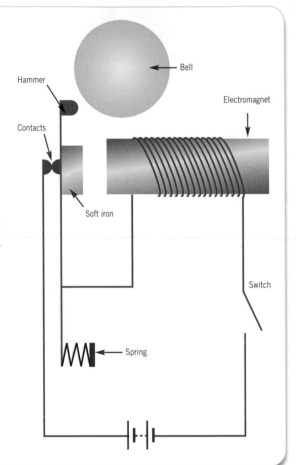

Hammer

Bell

Electromagnet

Contacts

Soft iron

Switch

Spring

Quick Test

1. What is the space around a magnet known as?
2. What is a compass and what does it do?
3. What is the name given to the tiny magnetic areas in a magnetic material?
4. What happens to a magnet if it's broken in two?
5. What is an electromagnet?

Key Words Exercise

Match each key word to its meaning.

Attract	If a current is passed through this it becomes an electromagnet
Circuit breaker	A force that pulls together
Coil	The area around a magnet
Core	This will make an electromagnet stronger
Domains	A force that pushes away
Electromagnet	A safety device that switches off with high current
Field lines	Tiny magnetic areas inside some metals
Magnetic field	These show the strength and direction of a magnetic field
Repel	This can be magnetised but it doesn't retain its magnetism
Soft iron	A magnet that can be switched on and off

Comprehension

Read the passage about smashing particles, then answer the following questions.

1. What does a particle accelerator do?

2. Why do the particles need to be accelerated to such high speeds?

3. What is the purpose of the magnetic field in the Cyclotron?

4. What is the name of the largest and most powerful particle accelerator in the world?

A particle accelerator accelerates particles at very, very high speeds. The particles are smashed into other particles so that they break up. Particle accelerators are used to discover new particles and their behaviour.

The Cyclotron is a circular accelerator that consists of two metal half circles in between the north pole and the south pole of a huge magnet. The magnetic field accelerates the particles round and round until they're going fast enough for a collision.

The world's largest and most powerful particle accelerator is in Cern, an international physics laboratory in Switzerland. The powerful Large Hadron Collider uses a magnet that's as big as a house and as heavy as the Eiffel Tower!

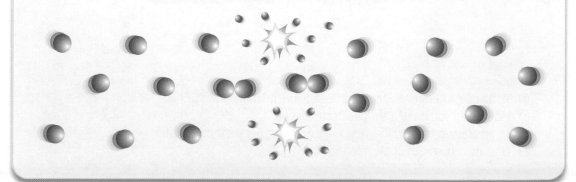

Magnetism and Electromagnetism

Testing Understanding

1 Use the following words to complete the sentences about magnetism and electromagnetism.

attract   circuit   contact switch   current   electromagnet

flow   high   reset   safety   spring

a) A circuit breaker is a _____ device that uses an _____ to switch off an electric current if it gets too _____.

b) The current flows through a _____ and an electromagnet. If the _____ gets too high, the electromagnet becomes strong enough to _____ a soft iron bar. This allows the contact switch to _____ upwards, breaking the _____ and switching off the current.

c) A circuit breaker can be _____ by pressing the reset button, allowing the current to _____ again.

2 Answer the questions below.

a) Draw the shape of the field using field lines around the permanent magnet below.

b) What can an electromagnet do that a permanent magnet can't do?
c) How could you make a simple electromagnet?
d) In an electric bell, explain how the electromagnet is switched off.
e) Give two other uses of electromagnets.

Lorna and Douglas want to investigate how the strength of an electromagnet will increase when they increase the number of turns of the coil.

They plan to do this by counting the number of paper clips that can be suspended by the electromagnet. They have made a simple electromagnet by wrapping insulated wire around an iron nail.

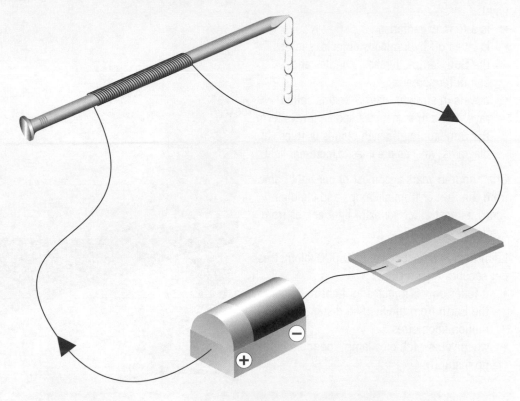

1. List the equipment that Lorna and Douglas will need to carry out the experiment.

2. Explain briefly how they should carry out the experiment.

3. The table below shows the results of their experiment. Plot the results on a graph and draw a curve through the points. Choose appropriate scales for your axes.

Number of Turns of the Coil	4	8	12	16	20	24
Number of Paper Clips Suspended	1	2	3	4	5	5

4. What is the relationship between the number of turns of the coil of the electromagnet and the number of paper clips that can be suspended?

5. Identify two sources of error in their experiment.

6. Suggest another way that they could make their electromagnet stronger.

Light Rays

Light

Light...
- is a form of radiation
- is given off by **luminous** objects, such as the Sun, lamps, torches, candles and computer screens
- travels in straight, narrow beams called rays. You can sometimes see light rays from the Sun shining through clouds or trees. In diagrams, you can use lines to represent light.

One hundred years ago, most of our light came from the Sun or from oil lamps and candles. Today, almost all of our extra light comes from electrical sources.

Light travels at a speed of 300 000 kilometres per second. As a result...
- it takes only 8 minutes for light to reach the Earth from the Sun – a distance of 150 million kilometres
- when you switch on a lamp, the room lights up instantly.

The Passage of Light

Some objects allow light to pass through them while others don't:
- **Transparent** objects, such as windows, allow light to pass through them.
- **Opaque** objects allow no light to pass through them, so they cause **shadows**. A shadow is an area where there's no light.
- Materials that allow some light through are known as **translucent** (e.g. stained glass).

The window opposite is transparent, so light can pass through it. The tree is opaque, so light can't pass through it and a shadow is formed.

How You See Things

Most objects that you see don't give off their own light. You see them because light rays from luminous objects bounce off them into your eyes. This is called **reflection**.

An object in a dark room can't be seen until the lamp is switched on. Then light shines onto the object from the lamp and reflects into your eyes. The Moon reflects light from the Sun, enabling you to see it at night, even when you can't see the Sun.

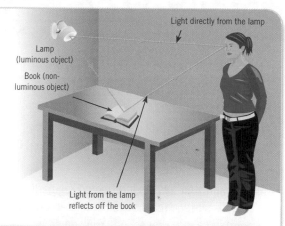

Light directly from the lamp

Lamp (luminous object)

Book (non-luminous object)

Light from the lamp reflects off the book

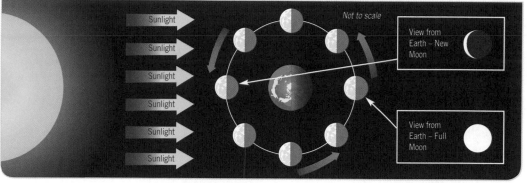

Sunlight

Not to scale

View from Earth – New Moon

View from Earth – Full Moon

Reflection

When light rays hit a surface, they're...
- reflected
- absorbed **or**
- transmitted through the material.

A rough surface (e.g. paper or wood) **scatters** the light so that it reflects in all directions.

Smooth surfaces (e.g. glass, mirrors and polished metal) reflect the light in a regular way so an image, or a reflection, can be seen.

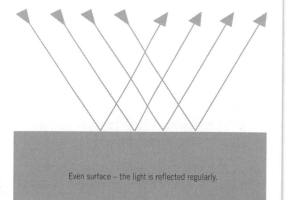

Uneven surface – the light is scattered.

Even surface – the light is reflected regularly.

Light Rays

Reflection and Images

A ray of light that hits a surface is called the **incident ray**. The angle of incidence is equal to the angle of reflection.

When light that has been reflected regularly by a smooth surface reaches your eyes, an image is formed called a reflection. You see an image because your brain doesn't see the light bend at the surface; it sees the light rays as if they're coming from inside the mirror.

A reflection is...

- the same size as the object
- the same distance from the mirror as the object
- laterally inverted (left becomes right and right becomes left)
- a **virtual** image (it can't be focussed on a screen; you can only see it by looking into the mirror).

Mirrors can help drivers to see around corners. They can also be used to make a periscope, to see around corners or over crowds.

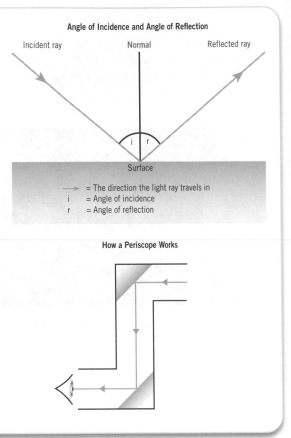

Angle of Incidence and Angle of Reflection

Incident ray   Normal   Reflected ray

i   r

Surface

→ = The direction the light ray travels in
i = Angle of incidence
r = Angle of reflection

How a Periscope Works

Refraction – Bending Light Rays

When light travels through transparent materials, such as water or glass, it travels slower than it travels in air or in a vacuum.

As a result...

- when light enters water or glass it bends (**refracts**) towards the normal
- when light leaves water or glass it speeds up and refracts away from the normal.

Because of refraction, water looks shallower than it really is and a pen standing in a glass of water looks like it bends at the surface.

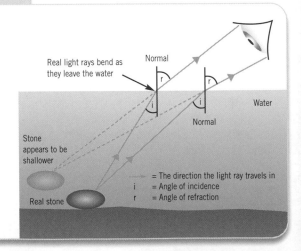

Real light rays bend as they leave the water

Normal

r   r

i   i   Water

Normal

Stone appears to be shallower

Real stone

→ = The direction the light ray travels in
i = Angle of incidence
r = Angle of refraction

Colour

When white light passes through a triangular prism made of glass or plastic, a spectrum is formed of all its constituent colours. Light behaves as if it travels in waves. White light contains light of different wavelengths and each wavelength is seen by your eyes as a different colour.

When the light passes through the prism, the different wavelengths are spread out and a spectrum is formed. This is because the light is refracted by the prism. The angle of refraction is different for each wavelength, so the different colours are seen. In the same way, a rainbow is formed by sunlight shining through droplets of rain.

Lighter coloured objects reflect more light than darker objects:

- White objects reflect all the light that hits them.
- Black objects absorb all the light that hits them.
- A red object absorbs light of all wavelengths except the wavelength that you see as red light, which is reflected.
- A blue object absorbs light of all wavelengths except the wavelength that you see as blue light, which is reflected.

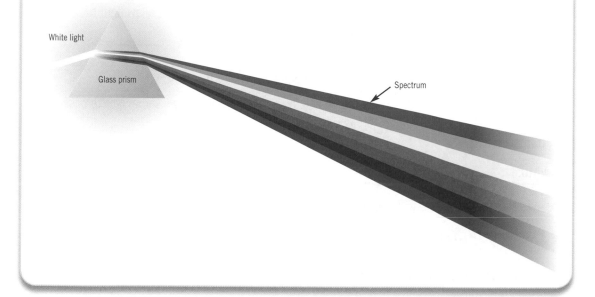

White light

Glass prism

Spectrum

Quick Test

1. What is a luminous object?
2. How long does it take light to reach the Earth from the Sun?
3. If light hits a rough surface, what happens to it?
4. Give two properties of a reflection in a mirror.
5. Why does water look shallower than it actually is?

Light Rays

Match each key word with its meaning.

Incident ray	The bending of light when it changes speed
Luminous	Light reflected in all directions
Normal	A line drawn at right angles to a surface at the point where the incident ray hits the surface
Opaque	Light rays that bounce off a surface
Reflection	An area where there's no light
Refraction	A word describing a material that doesn't allow light to pass through it
Scattering	A ray that strikes a surface
Shadow	An image that can't be focussed on a screen
Translucent	A word describing a material that allows some light to pass through it
Transparent	An object that gives out its own light
Virtual	A word describing a material that allows light to pass through it

Comprehension

Read the passage about light, then answer the following questions.

1. In the 1600s, why did most scientists prefer Newton's theory of light?

2. In the 1800s, Huygens's theory became more popular. Explain why.

3. How did Einstein change this belief in the early 1900s?

4. What is wave-particle duality?

In the 1600s, Christian Huygens, a Dutch physicist, claimed that light behaved as a wave, in a similar way to water or sound waves, travelling in straight lines. At about the same time, Isaac Newton put forward a theory that light travelled as tiny particles that moved through the air. Most scientists preferred Newton's ideas because he used his ideas to explain reflection and refraction, and he was well known for other impressive scientific ideas.

In the early 1800s, other scientists showed behaviour of light that could only be explained with wave ideas. Huygens's theory became more popular. However, in the early 1900s, Albert Einstein proved that light behaved as a particle, using evidence from an experiment carried out by Max Planck, a German physicist.

Today it's accepted that light can be thought of as two different things, a wave and a particle. Some of light's behaviour can be described with waves, and some with particles. This theory is known as wave-particle duality.

Testing Understanding

1 **Fill in the missing words to complete the sentences about light rays.**

a) Objects that give off their own light are known as _____ objects. You see other objects because light from luminous sources _____ off them into your _____ .

b) _____ materials allow light to pass through them, _____ materials don't allow any light to pass through them and _____ materials allow some light to pass through them.

c) Light travels in _____ lines. This means that an opaque object creates an area where there's _____ light, called a _____ .

2 **Answer the questions about reflection.**

The diagram below shows a ray of light striking a flat mirror.

a) Draw a normal at the point where the ray hits the mirror.
b) Label the angle of incidence.
c) Complete the ray to show the path of the reflected ray.
d) Label the angle of reflection.

Light Rays

Amy and Syama design an experiment to investigate how the angle of refraction changes for different materials. They have two blocks of different materials (glass and Perspex), some paper, a pencil, a ruler and a ray box.

1. Name a suitable instrument that can be used to measure the angles.

2. Syama suggests that they should repeat the experiment three times for each material. Suggest why this is a good idea.

3. Draw a diagram of a ray of light entering a block of glass. Label the normal, the incident ray, the refracted ray, the angle of incidence and the angle of refraction.

4. Amy suggests that they should repeat the experiment to find the angle of refraction when a ray of light enters a beaker of water. Give two reasons why the angle of refraction they measure will probably not be exactly the same as the true angle of refraction for water.

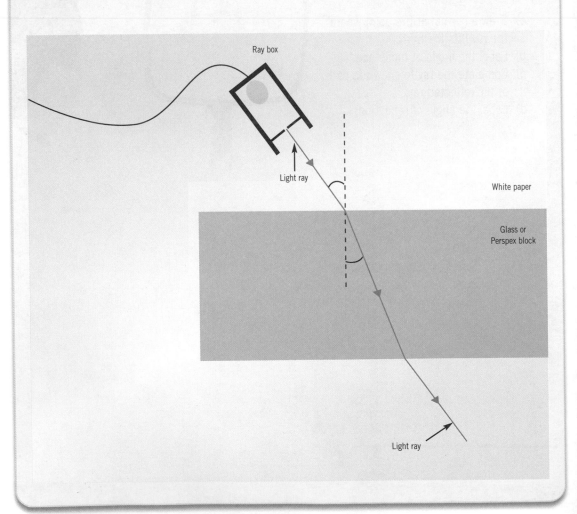

Ray box

Light ray

White paper

Glass or Perspex block

Light ray

Making Sound, Hearing Sound

Sound Waves

Sound, like light, is energy that travels in waves:
- Light waves are electromagnetic waves.
- Sound waves are mechanical waves.

Sound waves...
- are made by particles moving backwards and forwards in a wave motion
- carry energy from one place to another.

An object makes a sound wave when it vibrates:
- You can make a ruler vibrate over the edge of a desk by holding one end and applying a force to the other end and letting go.
- Musical instruments have vibrating parts – guitar strings vibrate and a drum skin vibrates.
- When you talk or sing, your vocal cords vibrate. You can feel them if you place your fingers gently against your throat.

Vibrations

A vibration is a repeating backwards and forwards motion of particles. As the particles vibrate, some are squashed together and others are spread apart.

Vibrations can travel through different materials, for example, through the air when you play a musical instrument:

- The instrument makes the air particles close to it vibrate and the vibrations are passed onto more air particles.
- When the vibrations reach your ears, you detect them as sound.

The air particles don't travel, they only vibrate. It's the sound waves that travel from the instrument to the ear.

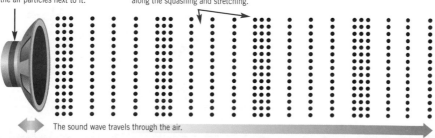

The loud speaker vibrates right and left, squashing and stretching the air particles next to it.

The air particles vibrate and in turn cause other air particles to vibrate, passing along the squashing and stretching.

The sound wave travels through the air.

The ear detects the vibrations of the particles as sound.

Making Sound, Hearing Sound

The Human Ear

Sound waves are directed into the ear canal by the outer ear:

- The sound waves make the ear drum vibrate.
- These vibrations are passed onto the **cochlea** by a set of three small bones called the **hammer**, anvil and stirrup.
- In the cochlea, a liquid moves backwards and forwards and stimulates nerve cells, which send impulses to the brain through the **auditory nerve**.

The ears detect the direction that a sound is coming from by sensing which ear is closer to the sound.

Owls listen carefully with both ears to identify where a sound is coming from. Then they turn their head to make another measurement. This is how they can precisely locate their prey.

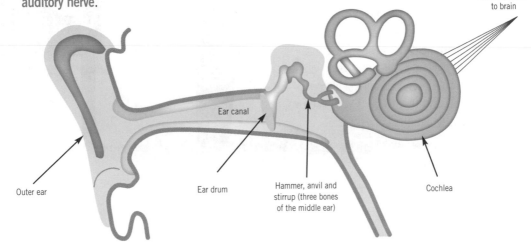

Auditory nerve to brain

Ear canal

Outer ear

Ear drum

Hammer, anvil and stirrup (three bones of the middle ear)

Cochlea

How Sound Travels

Sound travels through liquids and solids in the same way as it does through air, but it travels faster because the particles are closer together.

When you're in water, it's harder to tell where a sound is coming from because it travels faster and reaches both ears at virtually the same time.

Sound can't travel through a vacuum because there are no particles to vibrate.

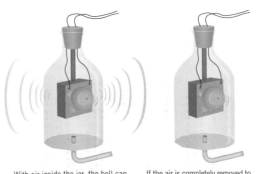

With air inside the jar, the bell can be heard and seen to be ringing.

If the air is completely removed to form a vacuum, the bell can be seen to be ringing but it can't be heard.

Frequency

The speed of the vibration of a sound wave is called the **frequency**:

- The unit of frequency is the Hertz or Hz.
- The frequency of a wave is the number of complete vibrations per second.
- Human ears can detect frequencies between 20Hz and 20 000Hz.
- The frequency affects the pitch of the sound. The higher the wave frequency, the higher the pitch.

Dolphins and whales can make sound waves of a wide range of frequencies, both lower and higher than your ears can detect. They use these sounds for hunting, navigating and communicating. Whales and dolphins communicate over many kilometres with very low frequency sound waves.

Using Sound Waves

Bats produce high frequency sounds, called ultrasound, in order to navigate. The bats detect the size and position of objects from the echoes produced by the sounds. The use of echoes to measure distances is called **echolocation**.

Echolocation can also be used on ships to measure the depth of water beneath them:

- A pulse of ultrasound is sent to the sea bed.
- The longer it takes for the echo to come back, the deeper the sea.

Ultrasound can also be used to...

- scan foetuses in the womb
- break up kidney stones without surgery
- clean surgical instruments.

A snake detects vibrations in the ground with the lower part of its jaw bone. The bone transmits the vibrations to its internal ears.

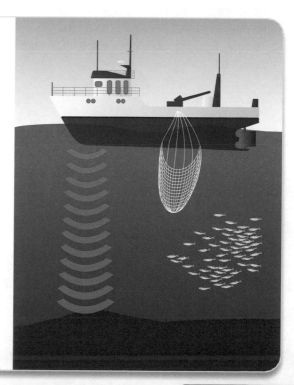

Making Sound, Hearing Sound

Describing a Wave

A sound wave vibrates particles to either side of their **rest position**. The distance the particles move each side, measured from their rest position, is called the **amplitude**.

You can't actually see sounds, but with a **microphone** connected to an **oscilloscope** you can see wave shapes on a screen. The microphone changes sound waves into electrical signals. The oscilloscope converts the signals into a moving wave shape on the screen called a waveform.

The **wavelength** is the distance from a point on a wave to the exact same point on the next wave (i.e. the length of one wave). High frequency sound waves correspond to high pitch notes and have shorter wavelengths.

The amplitude is the maximum distance the wave reaches from rest position. A sound wave with a large amplitude has a loud volume. A sound wave with a small amplitude is quieter.

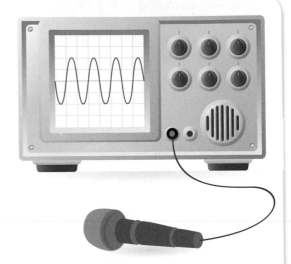

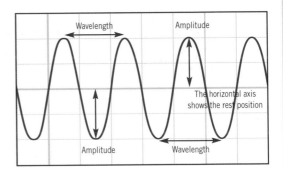

Wavelength

Amplitude

The horizontal axis shows the rest position

Amplitude

Wavelength

Quick Test

1. Sound waves are the vibration of particles. Do the particles move from one place to another?
2. What do sound waves carry from one place to another?
3. How does an owl identify where a sound is coming from?
4. Why can't sound travel through a vacuum?

Key Words Exercise

Match each key word to its meaning.

Amplitude	A coiled membrane full of liquid and nerve cells
Auditory nerve	The maximum distance from rest position
Cochlea	A sound wave can be viewed on one of these
Echolocation	This sends electrical impulses to the brain
Frequency	The number of complete waves per second
Hammer	A bone in the middle ear
Microphone	Measuring distances by detecting echoes
Oscilloscope	The normal position of a particle when it's not moving
Rest position	This converts sound energy into electricity
Wavelength	The distance between a point on a wave and the same point on the next wave

Comprehension

Read the passage about the dangers of noise, then answer the following questions.

1. What is the scientific name for a hole in the ear drum?

2. What permanent damage can be done inside the cochlea?

3. Name three environments where people can be exposed to loud noise.

4. How can people protect themselves from loud noises?

Exposure to a loud noise can cause damage to the ear drum, preventing it from vibrating efficiently, and hearing is lost. A hole in the ear drum, called a perforation, can sometimes heal and normal hearing can be restored.

However, if a person is exposed to loud noise over a long period of time, nerve endings in the cochlea can be damaged. This causes permanent loss of hearing. People who work in noisy environments, such as machinery workers, airport staff or members of music groups, are at risk from this kind of deafness. Ear protectors can be worn to reduce the amount of sound energy entering the ears.

Making Sound, Hearing Sound

Testing Understanding

1 **Fill in the missing words to complete the sentences about the human ear.**

a) Sound waves are directed into the ear _____ by the _____ ear.

b) Sound waves strike the ear _____, causing it to vibrate. These vibrations are passed onto the _____ by a set of three small bones called the hammer, anvil and _____.

c) In the cochlea, a liquid moves backwards and forwards and stimulates nerve _____ inside it that send _____ to the brain through the auditory nerve.

d) Your ears detect the _____ a sound is coming from by sensing which _____ is closer to the sound.

e) You may have noticed in _____ that it's harder to identify where a sound is coming from because the sound travels _____ and reaches both ears at almost the same _____.

2 **Read the information about frequencies, then answer the questions that follow.**

The highest frequency sound, or highest pitch note, that a human can detect is 20 000Hz. Other animals can detect sounds at much higher frequencies.

The graph opposite shows some of the maximum frequencies that different animals can hear.

a) What is meant by the frequency of a wave?

b) Which animals could hear a note of frequency 85 000Hz?

c) Add to the graph the maximum frequency a human can hear.

d) A dog whistle is used to call dogs. It can be heard by dogs but not by humans. Suggest a suitable frequency for the note produced by a dog whistle.

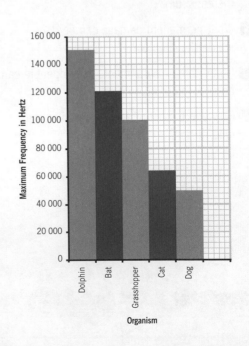

Mandy and Ivana want to use an oscilloscope to investigate different sounds. They want to compare the volume of different sounds with the amplitude of the waveform produced.

1 As well as an oscilloscope, what other piece of equipment do they need?

2 What is the amplitude of a wave?

Mandy has found a chart in a book that lists different sounds and their average loudness in decibels. She suggests that they could compare their results with the chart. The chart is shown below.

Sound	Loudness (decibels)
A whisper	25
Normal conversation	55
A hair dryer	60
A vacuum cleaner	80
Noisy traffic	80
A loud clap	90
A jet plane	125

3 Explain how Mandy and Ivana could use their equipment to measure the amplitude of a sound.

The results obtained in the experiment are recorded in the table below.

Sound	Amplitude on the Oscilloscope Screen (mm)
A whisper	10
Normal conversation	25
A hair dryer	25
A vacuum cleaner	35
Noisy traffic	45
A loud clap	60

4 Suggest how the girls could investigate whether there is a correlation between the sound levels in the chart and their results.

5 Carry out your suggestion to see whether there is a correlation or pattern between the two sets of results.

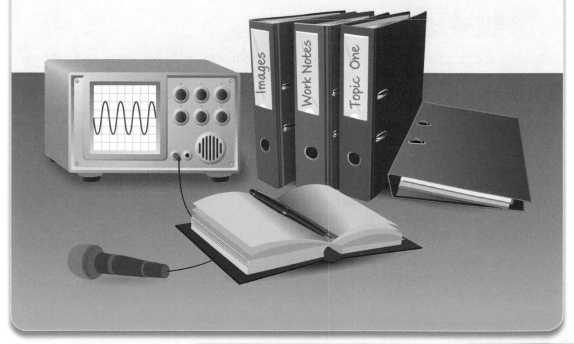

Energy and Electricity

Where Electrical Energy Comes From

Energy can be harnessed from...
- the wind
- the Sun
- water in rivers and reservoirs
- tides in the sea.

Wind and water are used to drive **turbines**, which turn **generators** to create electricity. Power stations also drive turbines using high pressure steam that's heated by fuels such as coal, gas, oil, nuclear fuel or biomass.

Some forms of energy can be stored for later use. Fuels, batteries and your body store chemical energy:
- Fuels transfer their chemical energy into light and heat when they're burned.
- Your body continuously transfers chemical energy into other forms in order to grow, to move and to keep warm.

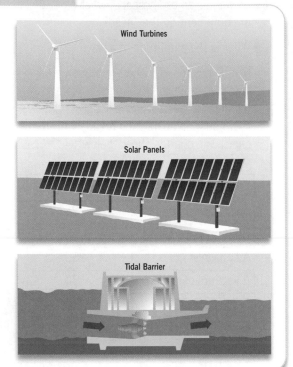

Wind Turbines

Solar Panels

Tidal Barrier

Measuring Electrical Energy

Batteries and cells transfer chemical energy into electrical energy when they're connected in a circuit:
- A device called a voltmeter can measure the energy being transferred from the cell to the electric current.
- A bulb in a circuit transfers electrical energy into heat and light.
- A second voltmeter can be used to measure the energy being transferred from the current to the bulb.

Voltmeters measure a quantity called **potential difference**. The unit of potential difference is volts (V).

The voltmeters in the circuit below both show a reading of 2.5V. The amount of energy transferred from the cell to the current is equal to the amount of energy transferred from the current to the bulb. This shows that energy is conserved.

2.5V

2.5V

Electrical Energy from Power Stations and Generators

Electricity is usually generated in a power station by rotating a coil of wire between the poles of a magnet:

- The rotation of the coil of wire between the poles of the magnet produces a current in the coil of wire. The magnet and coil of wire together are known as a generator because they generate or create electric current.
- The rotating coil is driven by a turbine that's blown around by high pressure steam. The steam is produced by water that's heated by burning fuel or by a nuclear reaction.

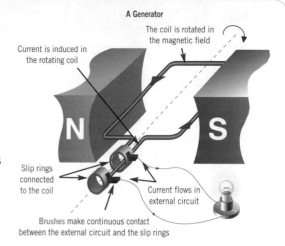

A Generator

The coil is rotated in the magnetic field

Current is induced in the rotating coil

Slip rings connected to the coil

Current flows in external circuit

Brushes make continuous contact between the external circuit and the slip rings

How a Thermal Power Station Works

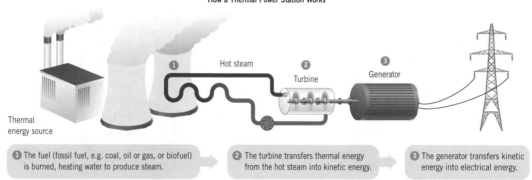

Hot steam

Turbine

Generator

Thermal energy source

1 The fuel (fossil fuel, e.g. coal, oil or gas, or biofuel) is burned, heating water to produce steam.

2 The turbine transfers thermal energy from the hot steam into kinetic energy.

3 The generator transfers kinetic energy into electrical energy.

Electrical Current

There are two types of electrical current:

- Generators produce a current that changes direction every time the magnet turns. This is called AC, or alternating current. AC is the current supplied to homes.

- Current supplied by a cell or battery is known as DC, or direct current.

AC and DC current can be displayed on a screen.

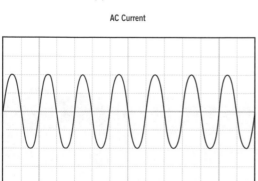

AC Current

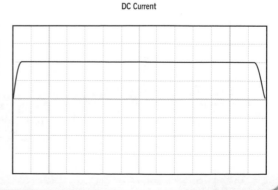

DC Current

Energy and Electricity

How Electrical Energy Reaches Your Home

Electricity that's generated in power stations is sent to towns and cities through a network of cables known as the **National Grid**:

- The National Grid cables are supported by pylons, or are buried underground. They carry electric current at very high voltages.
- It's more efficient to transmit electricity at a much higher voltage than is required by consumers. Increasing the voltage decreases the current in the transmission cables. The cables lose less energy to heat when the current is lower.
- The National Grid distributes electricity to consumers at different voltages using step-up and step-down **transformers** in the **substations**.
- Substations in cities, towns and villages reduce the voltage to a safer level for industrial and home use. Different consumers, such as factories, homes and schools, require different voltages. The **mains electricity** that comes to your home is at a voltage of 230–240V.
- Electricity meters measure the electrical energy used in each house. Wires inside the walls carry the current to the electric sockets and lights.

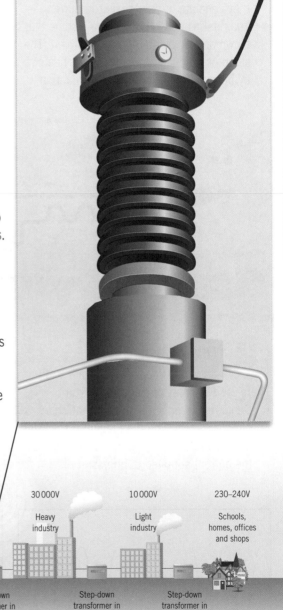

A Transformer

Distributing Electricity

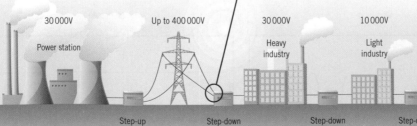

30 000V	Up to 400 000V	30 000V	10 000V	230–240V
Power station		Heavy industry	Light industry	Schools, homes, offices and shops
	Step-up transformer in a substation	Step-down transformer in a substation	Step-down transformer in a substation	Step-down transformer in a substation

Mains Electricity

Anything that transfers energy from one form to another is called a **device**:

- Devices that use electrical energy include toasters, hair dryers, fans, computers, kettles and DVD players.
- Whenever you plug in a device, you're connecting it into an electrical circuit.

In the UK, three-pin plugs are used to connect a device to a mains circuit. Plugs contain a fuse and each pin is connected to a wire:

- The live wire controls the AC current, flowing backwards and forwards many times per second.
- The neutral wire completes the circuit.
- The earth wire is a safety wire that connects the metal casing of a device to the earth. The earth wire works with the fuse to protect anyone who uses the device. If a loose wire inside the device touches the metal casing, a large current immediately flows through the earth wire and the fuse blows. This disconnects the circuit, preventing a possible electric shock or overheating that could cause a fire.

Most modern devices are made of plastic. This adds another layer of insulation because plastic is an electrical insulator. This is called **double insulation**. Devices with double insulation don't need an earth wire connected to the casing.

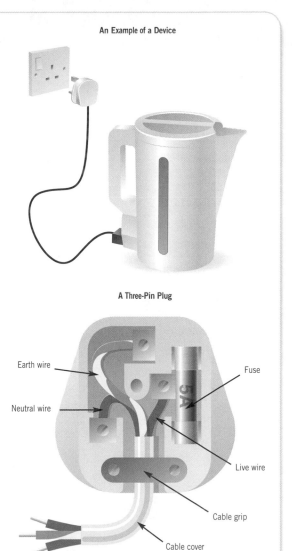

An Example of a Device

A Three-Pin Plug

Earth wire

Neutral wire

Fuse

Live wire

Cable grip

Cable cover

Quick Test

1. Name three different sources of energy.
2. Name three ways that energy can be stored for later use.
3. What is the unit of potential difference?
4. What is DC current?
5. What is AC current?

Energy and Electricity

Match each key word to its meaning.

Alternating current	A coil of wire rotating between the poles of a magnet
Brushes	This usually drives a generator in a power station
Device	This electricity comes to your home at 230–240V
Direct current	A device with a plastic casing
Double insulated	This changes the voltage to a safer level for use in homes
Generator	This tells you how much energy is transferred in a cell or in a bulb in an electric circuit
Mains electricity	A network of cables taking electricity from power stations to homes
National Grid	Something that transfers energy from one form to another
Potential difference	The parts of the National Grid that contain step-up or step-down transformers
Substations	Current that continuously changes direction, created by a generator
Transformer	These ensure continuous contact of the wires in a generator as the coil rotates
Turbine	Current supplied by a cell or battery

Comprehension

Read the passage about the waste products of electricity generation, then answer the questions.

1. Describe an effect of a polluting gas on the environment.

2. Give three disadvantages of wind turbines.

3. a) Give one advantage of using nuclear fuel to produce electricity.
   b) Give one disadvantage of using nuclear fuel to produce electricity.

Burning fossil fuels, such as oil, coal and gas, always produces waste. Waste gases can pollute the air, soil and water. Some polluting gases, such as carbon dioxide, increase the greenhouse effect and lead to global warming.

Renewable sources produce less pollution but can create other problems for the environment. Wind turbines change the landscape and produce loud noises that can disturb wildlife. Wind turbines only generate a small amount of electricity – one power station running on solid fuel generates about 2000 times more energy than a wind turbine. Also, another source of energy is required when the wind isn't blowing.

Nuclear fuel is a very efficient way of producing electrical energy and releases no harmful gases into the atmosphere. However, nuclear waste material needs to be disposed of safely, which can be very expensive. Many people also worry about the risk of accidental emission of radioactive material.

1 **Fill in the missing words to complete the sentences about energy and electricity.**

a) Electricity is generated in _____ and sent to towns

and cities through a network of _____ known as the

_____ . The cables are supported by

_____ , or are buried underground. They carry electric current at very

_____ voltages.

b) It's more efficient to transmit _____ at much higher voltages than is

required by _____ and factories. Increasing the voltage decreases the

_____ in the transmission cables. The cables lose _____

energy to _____ when the current is lower.

c) Step-up and _____ transformers are used at _____ to

reduce the voltage to a _____ level for industrial and home use. The

_____ electricity that we use in our homes is at a voltage of 230–240V.

2 **The diagram below shows a correctly wired three-pin plug.**

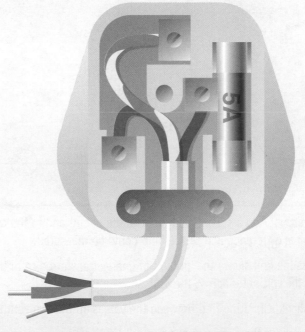

a) Label the live wire, the neutral wire, the earth wire and the fuse in the diagram.
b) Describe and explain the functions of the live wire and the neutral wire.
c) Many domestic devices are double insulated. What does this mean?
d) Which wire in a three-pin plug isn't required if a device is double insulated?

Energy and Electricity

Christine wants to investigate how the energy supplied by cells is used in an electric circuit. She set up the circuit below to measure the potential difference across a cell with a voltmeter in a simple circuit.

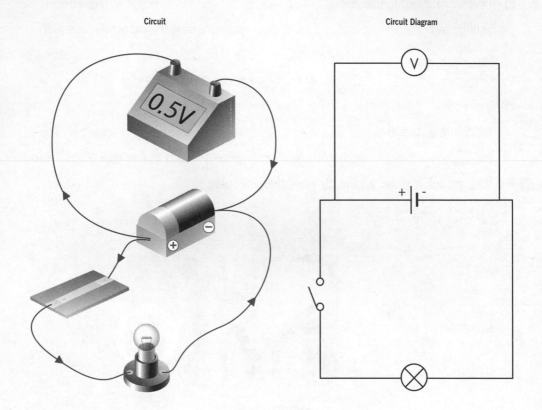

Circuit

Circuit Diagram

1 Copy the diagrams above. Add a second voltmeter to the circuit and to the circuit diagram to show how the potential difference across the bulb could be measured.

2 The voltmeter across the cell shows the amount of energy transferred from the cell to the current. What does the voltmeter across the bulb show?

3 What relationship should Christine find between the values on the two voltmeters?

4 What does this show about the energy in the circuit?

5 Christine adds a second cell next to the first cell in series. Draw a circuit diagram that shows how she can investigate whether energy is conserved in this circuit.

6 How would you expect the values of the potential difference to change?

Pushing and Turning

Pushing Forces

When a force pushes on a surface, it causes **pressure**. The amount of pressure depends on...
- the size of the force applied
- the area that the force is acting on.

If a force is spread over a larger area, the pressure is less. If a force is concentrated on a smaller area, the pressure is greater.

For example, the pressure caused by a woman treading on your toe wearing trainers would be less (and it would hurt less!) than the same woman treading on your toe if she was wearing stiletto heels. This is because, although the size of the force is the same, the area of the stiletto heel is much smaller.

Increasing Pressure

The following examples show how pressure can be increased by reducing the area over which a force acts:
- Studs on football boots provide better grip.
- A knife has a sharp edge with a very small area, allowing it to cut easily.
- When you push a drawing pin, the force is spread over the area of the head of the pin. The same force is concentrated over the much smaller area of the pin point, creating much more pressure. As a result, the pin sticks into the notice board, not into your finger.

Pushing and Turning

Decreasing Pressure

The following examples show how pressure can be decreased by increasing the area over which a force acts:
- Snow shoes have a large area to stop the person wearing them sinking into the snow.
- Camels' feet have a large area to prevent them from sinking into the sand.

Calculating Pressure

The pressure exerted by a force is calculated by:

$$\text{pressure} = \frac{\text{force}}{\text{area}}$$

The unit of pressure is newtons per square metre (N/m^2) or newtons per square centimetre (N/cm^2).

For example, you could calculate the pressure exerted by a metal cube of weight 500N with dimensions 10cm × 10cm × 10cm:

$$\text{pressure} = \frac{\text{force}}{\text{area}} = \frac{500N}{100cm^2} = 5N/cm^2$$

Pressure in Gases

Gases exert pressure as the gas particles **collide** with the walls of the container they're in.

The gas particles inside a balloon exert pressure on the walls of the balloon. This allows the balloon to hold its shape.

There are different ways of increasing the pressure in a gas:
- Increase the number of gas particles in a container of fixed volume. This means the particles collide with the walls more frequently.
- Reduce the volume of the container for a fixed number of gas particles. This means the particles collide with the walls more frequently.
- Increase the temperature of the gas to give the particles more kinetic energy. This means they collide with the walls more frequently and with a greater force.

Not to scale

The air around you is at **atmospheric** pressure:
- Normal atmospheric pressure is $101\,000N/m^2$.
- The atmosphere doesn't crush you because the pressure of the blood in your body is strong enough to balance the atmospheric pressure.

Pressure in Liquids

Unlike gases, liquids can't be compressed. But liquids can transmit pressure. **Hydraulic** systems use liquids to transmit pressure:

- Hydraulic brakes in a car use fluid to transmit pressure from the foot pedal to the brake pads.
- If an air bubble gets into the brake fluid, it can stop the transmission of the pressure, so the brakes may not work properly.

The pressure in liquids increases with depth:

- The pressure at the bottom of the ocean is very high because of the weight of the water pushing down. Deep-sea submarines and creatures that live at the bottom of the oceans must be able to withstand very high pressures.
- **Dams** are wider at the base because this is where the pressure of the water is greatest.

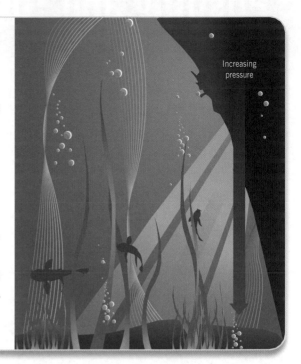

Increasing pressure

Turning Forces

A force can turn an object around a hinge or **pivot**. A **fulcrum** is another name for a hinge or pivot.

When you open a door, you're using a turning force. The door turns around the hinges. This turning force is known as the **moment** of the force.

The moment of a force depends on...

- the size of the force
- the distance between the force and the pivot.

A greater distance between the force and the pivot increases the moment of the force.

A door handle is as far away as possible from the hinges in order to increase the moment of the force. If the handle was close to the hinges, it would be difficult to open the door.

Other everyday objects also use the principle of moments:

- A spanner provides a moment to undo a nut. A spanner with a longer handle increases the moment of the force and can undo a tighter nut.
- A **lever** can be used to lift a heavy load or open a can of paint. Increasing the length of the lever increases the moment of the force.
- The wheel of a wheelbarrow acts as a pivot. The handles are placed as far away as possible from the wheel in order to increase the moment and make it easier to lift the load.

Hinge

Hinges

Door handle

Door handle

Hinge

Door is easy to open

Door is difficult to open

A long distance between the handles and the pivot makes the load easier to lift

Pushing and Turning

Balancing Moments

The two children on the seesaw below are **balanced**. The boy has a greater weight than the girl, but he is closer to the pivot.

It's possible to calculate the moment of each child using the following equation:

the moment of a force	=	force	×	distance from the pivot

The moment of the force created by the boy is equal to 900N x 2.0m = 1800Nm.

The moment of the force created by the girl is equal to 600N x 3.0m = 1800Nm.

The moments of the forces on each side of the seesaw are equal, so the seesaw balances.

Quick Test

1. What is the unit of pressure?
2. What two things does pressure depend on?
3. What two things does the moment of a force depend on?
4. Calculate the moment of a force of 4N that is 2.5m from the pivot.
5. Calculate the moment of a force of 4N that is 1.5m from the pivot.

Key Words Exercise

Match each key word to its meaning.

Atmospheric	This is equal to force / area
Balanced	The turning effect of a force
Collide	The point from which a turning force acts
Dam	Another word for pivot
Fulcrum	The pressure of the air around us
Hydraulic	This is how gas particles exert pressure on the walls of a container
Lever	A system that transmits pressure through a liquid
Moment	When the moments of the forces are equal
Pivot	This uses the principle of moments to lift a heavy load
Pressure	This is wider at the bottom because of the increase in the pressure of the water at a greater depth

Comprehension

Read the passage about levers, then answer the following questions.

1. Explain the connection between the word 'lever' and its original meaning in French.

2. Give four examples of levers that people regularly use.

3. What feature of levers allowed Archimedes to claim that he could use a lever to move the whole world?

4. Suggest why many historians believe that the ancient Egyptians used levers.

The word lever comes from the French word 'lever', which means 'to lift'. A lever uses a pivot to multiply a force applied to an object. You may not realise it, but levers are everywhere. The average person uses a lever many times each day. Imagine life without door handles, staplers, bike gears, car jacks, can openers and scissors. Almost every moveable joint is connected to a lever.

The ancient Greek, Archimedes, was the first to explain levers, describing a relationship between the force and the distance from the pivot. He once famously said 'Give me a lever long enough and a fulcrum on which to place it, and I shall move the world'. Many historians believe that ancient Egyptian construction techniques involved levers. This could explain how they managed to move the huge stone blocks they used to build the pyramids.

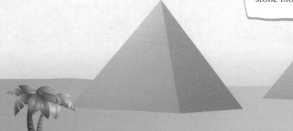

Pushing and Turning

Testing Understanding

1 **Fill in the missing words to complete the sentences about pushing and turning.**

a) If a _____ is spread over a larger area, the _____ is less.

If a force is concentrated on a smaller _____, the pressure is greater.

b) Pressure can be _____ by decreasing the area on which a force acts.

_____ on football boots increase the pressure because

_____ area is in contact with the _____ compared with

a normal trainer. This is why they give more grip. A _____ has a sharp

edge with a very small area, allowing it to cut easily.

c) Pressure can be decreased by _____ the area over which a force acts.

Snow shoes have a large area to _____ the pressure, preventing the

wearer from _____ into the snow. _____ feet have a

large area that decreases the pressure, preventing them from sinking into the sand.

2 **Look at the diagram below of two children sitting on a seesaw.**

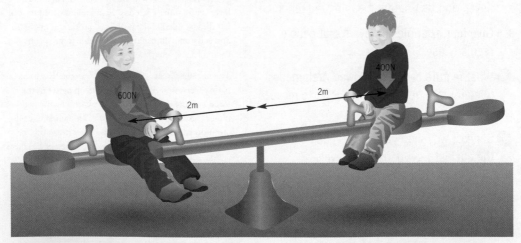

a) Calculate the moment created by the girl.

b) Calculate the moment created by the boy.

c) Explain why the seesaw isn't balanced.

d) Explain how the seesaw could be made to balance by moving the position of one of
the children.

Skills Practice

Selma and Bezede decide to investigate how the pressure of a gas varies as the gas is heated.

They have some gas sealed in a test tube with a rubber bung.

They use a pressure sensor and a temperature sensor connected to a data logger to measure the pressure in the gas as the temperature rises.

They set up some equipment as shown below.

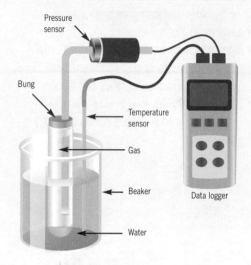

① Explain why it's better to heat the gas using water in a beaker rather than apply heat directly to the gas using a Bunsen burner.

② Selma predicts that the temperature of the gas will increase from room temperature to 90°C in about 6 minutes. Selma and Bezede need to program the data logger to take readings at suitable time intervals. Suggest how frequently the data logger should be programmed to take readings.

③ The results of the experiment are recorded in the table below. Carefully choose a suitable scale and plot a graph of pressure (y-axis) against temperature (x-axis) for the gas.

Time (s)	Pressure (N/m^2)	Temperature (°C)
0	101 000	20.0
30	102 300	23.5
60	103 600	27.0
90	104 600	30.5
120	104 900	34.0
150	106 200	37.5
180	107 500	41.0
210	108 800	44.5
240	110 100	48.0
270	111 400	51.5
300	112 700	55.0
330	114 000	58.5
360	115 300	62.0
390	116 600	65.5
420	117 900	69.0
450	119 200	72.5
480	120 500	76.0
510	121 800	79.5
540	123 100	83.0
570	124 400	86.5
600	125 700	90.0

④ Explain, in words, the relationship between the pressure and the temperature of the gas.

Speeding Up, Slowing Down

Speed

When you want to describe how fast something is moving, you measure its **speed**. Speed is a measure of how far an object travels in a specific time.

To determine the speed of an object, you need two pieces of information:
- The distance it has travelled.
- The time it has taken.

Speed is calculated using this equation:

$$\text{speed} = \frac{\text{distance}}{\text{time}}$$

The cyclist below travels between the two lamp posts in 5.0 seconds. The distance between the lamp posts is 100 metres.

The speed is equal to $\dfrac{100\text{m}}{5.0\text{s}} = 20\text{m/s}$.

This calculation gives the average speed of the cyclist since his actual speed may vary over the 5.0 seconds.

Some measuring instruments are designed for specific jobs and can give a direct measurement of speed:
- A policeman uses a radar gun to obtain measurements of distance and time. The radar gun then calculates the speed.
- An anemometer is a device designed specifically to measure the speed of the wind.

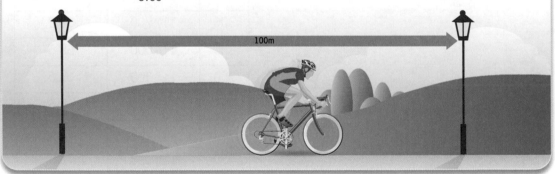

100m

Units

Metres per second (m/s) is a commonly used **unit** for speed, but sometimes different units are used:
- Road signs in Britain use the unit miles per hour (mph) but most other European countries use kilometres per hour (km/h).
- It would be easier to measure the speed of a snail in a smaller unit, such as centimetres per second (cm/s).

All of the units for speed are a distance unit divided by a time unit. If you want to compare the speeds of two different objects, you must use the same unit of speed for both objects.

Velocity

Velocity is a speed in a particular direction. The two cars below have the same speed, but different velocities.

20m/s to the left

20m/s to the right

Force and Velocity

A force can change the velocity of an object by...
- making an object go faster
- making an object go slower
- making a stationary object move
- making a moving object stop
- changing the direction of a moving object.

The acceleration of an object is defined as when the velocity changes. All of the objects below experience acceleration:
- A car speeds up as the engine drives it forwards.
- A ball stops as the goalkeeper catches it.
- A ball slows down as the force of friction acts on it.
- The Earth changes direction as it rotates in a circular orbit around the Sun.

On Earth, the force of gravity and usually the force of friction are acting on objects. If there were no forces acting on an object...
- a stationary object would remain stationary
- a moving object would continue to move at the same velocity (i.e. at the same speed in the same direction).

An ice skater experiences very little friction. This is as close as you can get on Earth to an object not having any forces acting on it. The ice skater glides a long way at the same velocity before she finally comes to rest.

Speeding Up, Slowing Down

Friction and Streamlining

Without friction...
- you wouldn't be able to walk
- wheels wouldn't grip on the road.

Brakes rely on friction to slow things down, producing heat energy. Moving parts in engines are oiled to reduce friction so that moving parts don't overheat. Skiers wax their skis to reduce friction, enabling them to go faster.

As objects travel through gases or liquids, frictional forces slow them down:
- These frictional forces are also known as **drag forces**.
- The effect of these forces can be reduced if an object is **streamlined**. A dolphin experiences a small drag force because its shape is very streamlined.

Balanced and Unbalanced Forces

A car engine needs to supply a driving force greater than the drag force, otherwise the car will slow down:

- When the driving force is greater than the drag force, the forces are **unbalanced** and the car accelerates.

- When the drag force is greater than the driving force, the forces are unbalanced and the car slows down.

- When the driving force and the drag force are equal, the forces are **balanced** and the car continues to move at a constant velocity.

A skydiver accelerates under the force of gravity (weight). This force is a downward force:
- As the skydiver falls faster, the upward drag force increases.
- Eventually, the drag force is great enough to balance his weight and the forces are balanced.
- He then falls at a constant speed called **terminal velocity**.

Air resistance

Weight

❶ His weight is greater than the drag force, so he accelerates. He has not yet reached maximum speed.

Air resistance

Weight

❷ His weight is equal to the drag force, so the forces are balanced. He has now reached maximum speed and his velocity is constant.

Describing Motion With Graphs

A distance–time graph illustrates the motion of an object. The **gradient** of the graph is equal to distance divided by time, which is equal to the velocity.

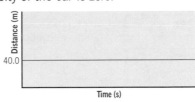

$$\text{gradient} = \text{velocity} = \frac{\text{distance travelled}}{\text{time}}$$

Example	Distance–time Graph
The car is stationary, 40.0m away from a tree.  Distance = 40.0m	The gradient of the line is zero because the velocity of the car is zero. 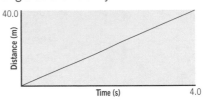
The car is travelling away from the tree. It takes 4.0s to travel 40.0m. Time in seconds	The distance–time graph has a gradient of 10.0 meaning that the velocity of the car is 10.0m/s. 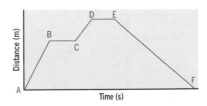
A cyclist travels at a constant speed from point A to B. She stops between B and C, then continues at a constant speed to D. The cyclist stops again between points D and E. Between E and F, the cyclist travels at a slower constant speed in the opposite direction, ending up at the starting position when she reaches point F.	The gradient changes as the velocity changes.

Quick Test

1. What two pieces of information are needed to calculate speed?
2. Calculate the speed of a car that moves 150 metres in 6.0 seconds.
3. Give two examples of a force changing the velocity of an object.
4. Give two examples of an object accelerating.

Speeding Up, Slowing Down

Key Words Exercise

Match each key word with its meaning.

Acceleration	This must be the same when comparing the speeds of two different objects
Balanced	An object that's shaped so that the drag force is less
Drag force	Equal to velocity on a distance–time graph
Friction	A change in velocity
Gradient	The distance an object moves in a certain time
Speed	Speed in a specific direction
Stationary	The maximum velocity reached by a skydiver when the drag force balances the weight
Streamlined	An object that's not moving
Terminal velocity	The force experienced by a moving object in a gas or liquid that opposes its motion
Unbalanced	A force that opposes motion and produces heat energy
Unit	When two forces in opposite directions are equal
Velocity	When two forces in opposite directions aren't equal

Comprehension

Read the passage about drag forces, then answer the following questions.

1. In which direction does a drag force act?

2. What effect does a greater speed have on the drag force?

3. How does a dolphin's body allow it to travel faster?

4. What can be learned from the study of creatures such as dolphins and sharks?

Whenever an object moves through water, it experiences a force called drag, opposing its motion. A stronger drag force means that a greater driving force is required in order to make the object accelerate. Drag forces increase if an object travels at a greater speed.

Dolphins and sharks have pointed noses and torpedo-shaped bodies that are larger at the front than they are at the back. The streamlined shape of their bodies allows them to travel faster though water because the drag force is reduced.

Researchers study the shape of naturally-streamlined creatures. Their research allows them to design boats, cruise liners and submarines that can travel faster and more efficiently.

Testing Understanding

1 **Fill in the missing words to complete the sentences about speeding up and slowing down.**

a) Speed is a measure of how far an object travels in a specific _____. To determine the speed of an object, you need to know the _____ it has travelled and the time that it took.

b) Velocity is speed in a specific _____. Both speed and velocity can be measured in the _____ metres per _____ (m/s).

c) _____ means a change in velocity. A _____ is required to change a velocity. It can change the velocity of an object by changing its _____ or by changing its _____.

2 **Study the information about a skydiver, then answer the questions that follow.**

The graph below shows the motion of a skydiver as she jumps out of a hot-air balloon for the first 80 seconds. She doesn't open her parachute during this time.

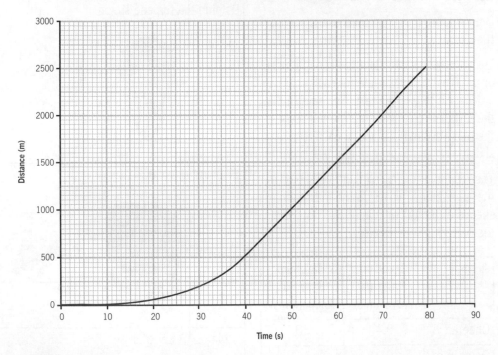

a) The gradient of this distance–time graph starts off at almost zero and then increases over the first 40 seconds of motion. What does this tell you about the velocity of the skydiver over this time?

b) During the period 0–40 seconds, what can you say about the forces on the skydiver?

c) i) Calculate the gradient of the graph between 40s and 80s.
   ii) What does this tell you about the velocity of the skydiver between 40s and 80s?

d) What can you say about the forces on the skydiver between 40s and 80s?

Speeding Up, Slowing Down

Skills Practice

Tulsy and Ben want to investigate whether there's a relationship between the speed of an animal and its shape.

They have made some shapes from Plasticine and will test how streamlined they are when falling through wallpaper paste.

The shapes they will use to represent each animal are shown in the table opposite.

1. What other piece of equipment do Tulsy and Ben need in order to carry out the experiment?

2. Draw a table that they could use to record their results.

3. Tulsy thinks it's important to keep the mass of the Plasticine the same for each shape in order for it to be a fair test. Ben says that as the different creatures are different sizes, the mass of the Plasticine in each test should change. Explain which idea is better.

4. The results of the experiment are shown in the table below, alongside the maximum speeds of the different animals. Describe the pattern in the results.

Animal	Speed of Plasticine Falling (m/s)	Maximum Speed of Animal (m/s)
Cheetah	0.4	30
Dolphin	0.2	16
Elephant	0.1	10
Eagle	0.3	20

5. Suggest three ways in which this experimental method doesn't model the real-life situation.

Animal	Shape to Represent the Animal
Cheetah	A triangle with a narrow front
Dolphin	A torpedo shape
Elephant	A cube
Eagle	A flat triangle

For further practice, go to p.36-39 of the Year 9 Science Essentials Workbook

Space and Gravity

Observing the Sky

Early observations of the sky helped people to understand time:

- Ancient civilisations studied the movement of the Sun, Moon and stars across the sky. These observations were used to measure time and the seasons, helping them to plan the planting and harvesting of crops.
- The Egyptians recognised fixed patterns of stars that appeared to move across the sky. By about 4000BC, they had developed the 365-day calendar.
- By about 1000BC, the Babylonians could predict the behaviour of the Moon and the planets in relation to distant constellations. Stonehenge was built in England, which lines up with the Sun's positions throughout the year.

The Geocentric Model

In 400BC, Aristotle realised that the Earth was a sphere and that different stars could be seen from different positions on the Earth's surface:

- Soon, the Pole Star, which is visible from all countries in the Northern Hemisphere, was being used to guide ships at sea.
- Aristotle wrongly believed that the Earth was the centre of the Universe and that the Sun, Moon, planets and stars moved around the Earth in circular orbits, supported by crystal spheres. This is known as a **geocentric** model.
- In 120AD, an Egyptian astronomer named Ptolemy developed a model of the Solar System, explaining Aristotle's ideas and agreeing that the Earth was at the centre.

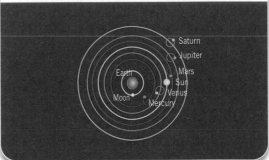

The Heliocentric Model

It wasn't until 1543 that Copernicus, from Poland, described a model for the Solar System with the Sun at the centre – a **heliocentric** model.

Church leaders of the time wouldn't accept the heliocentric model because they believed that God had put the Earth at the centre of the Universe.

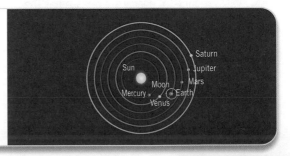

Space and Gravity

Improving the Model

When telescopes became more available, ideas began to change:

- In 1609, Kepler made observations and adapted Copernicus's model, changing the circular orbits of the planets around the Sun to elliptical orbits. Ellipses are squashed circles. This led to a greater understanding of how the speed of the planets changes as they orbit.
- There have been many other famous astronomers since, including Galileo and Hubble. The Hubble telescope has helped us to understand more about the movement of stars and galaxies in our Universe.

- Astronomers are still improving and adapting the model today as more information about our Universe is gathered.

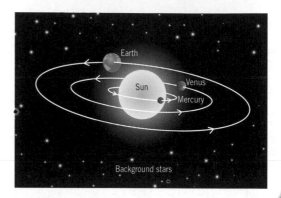

The Big Bang

Scientists today believe that the Universe began with a huge explosion called the **Big Bang**:

- In this explosion, energy, matter and forces came into being. Seconds later, everything began to cool and spread out.

- At this time, the force of gravity pulled together atoms of **hydrogen** and **helium** into clouds of gas and dust called **nebula**.
- Gravity pulled more gases together into stars, and pulled groups of stars into galaxies.

Inside a Star

Inside a star, the force of gravity pulls the gases together until the temperature and pressure are so great in the core that nuclear reactions occur:

- Vast amounts of energy are released as the star changes hydrogen into helium. Scientists have predicted that the Sun (a star) has enough hydrogen to keep it burning for 6000 million years!
- When a star's core has used up all of the available hydrogen fuel, the star's outer layer cools to a huge red ball called a **red giant**.

- Eventually the outer layer drifts away leaving a hot, dense core called a **white dwarf**. Later, the white dwarf also fades away as the star dies.
- Some stars are huge and instead of becoming a red giant, they explode. This explosion is known as a **super nova** and leaves behind a very small, dense **neutron star**.
- If an even larger star explodes, the gravity can be so strong that a **black hole** is formed. The gravity of a black hole is so strong that not even light can escape!

Gravity on Earth

Gravity is a force that acts between all objects that have mass. The force of gravity between two objects depends on...

- the mass of the objects (the greater their masses, the greater the force of gravity between them)
- the distance between the objects.

The force of gravity between two apples is too small to notice because apples have a small mass. The force of gravity between two planets is large because the planets are huge.

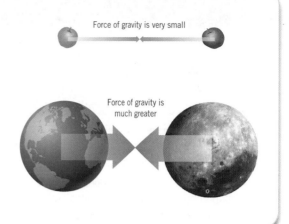

Force of gravity is very small

Force of gravity is much greater

Gravitational Force and Weight

The gravitational force between an object and the Earth is known as the weight of the object:

- On Earth, the force of gravity on an object (its weight) pulls it down.
- The weight of an object can be calculated using the gravitational field strength, which on Earth is equal to 10N/kg.

If something falls, it will fall with an acceleration caused by the force of gravity. This is called the acceleration of free fall.

It's said that Isaac Newton (1642–1727) saw an apple fall from a tree and accelerate under the force of gravity. He connected this force with the force that pulls the Moon towards Earth, keeping it in its orbit:

- Newton imagined the Moon to be continuously falling towards Earth, but never reaching it.
- He reasoned that the force of gravity kept the Moon falling and calculated how fast it would have to fall in order to keep orbiting the Earth, but to never reach it. When Newton made actual measurements of the motion of the Moon, he found that they matched his predictions.
- Newton showed that the force of gravity is what keeps objects in the Solar System in their orbits.

The Moon is pulled towards the Earth because it feels the gravitational field strength of the Earth. The Moon's gravity is felt weakly on Earth, but it's strong enough to affect the seas and oceans.

Tides are caused by the gravitational pull of the Moon. The gravitational pull of the Sun also affects the tides.

Space and Gravity

Satellites in Orbit

A force is required to change the direction of a moving object, such as the Earth rotating around the Sun. These forces are gravitational forces, and they hold planets and moons in their orbits:

- Planets close to the Sun experience large gravitational forces and their paths are curved.
- Planets further from the Sun experience weaker gravitational forces and follow less curved paths.

In the same way, an athlete has to exert a large inward force on a hammer to keep it going around in a circle. When he lets go, the force is no longer applied and the hammer moves away along a straight line.

Artificial satellites keep moving around the Earth without power:

- The only thing holding up satellites is the force of gravity from the Earth.
- A launch rocket takes the satellites into space and sets them into motion at the correct height and speed. A high orbit has a less curved path than a low orbit, so its speed can be less.

A Comet's Orbit

A comet has a highly-elliptical orbit around the Sun:

- When the comet is closest to the Sun, the gravitational force of the Sun on it is greatest and the comet speeds up.
- When the comet is furthest from the Sun, the gravitational force on it is weakest and the comet travels more slowly.

Quick Test

1. Who was the first person to suggest a heliocentric model?
2. Who opposed the heliocentric model and why did they oppose it?
3. What two things does the force of gravity between two objects depend on?
4. Why is the force of gravity between two apples very small?

KEY WORDS
Make sure you understand these words before moving on!

- Big Bang
- Black hole
- Geocentric
- Heliocentric
- Helium
- Hydrogen
- Nebula
- Neutron star
- Red giant
- Super nova
- White dwarf

Key Words Exercise

Match each key word with its meaning.

Big Bang	• •	A model of the Solar System with the Sun at the centre
Black hole	• •	A model of the Solar System with the Earth at the centre
Geocentric	• •	The gravity of this is so strong that not even light can escape
Heliocentric	• •	The fuel of a star
Helium	• •	A massive star dies in a big explosion known as this
Hydrogen	• •	Scientists' model for the beginning of the Universe
Nebula	• •	A star that has run out of fuel becomes this
Neutron star	• •	The small, dense core of a dying star
Red giant	• •	The dense core of a star left after a super nova
Super nova	• •	A cloud of gas and dust
White dwarf	• •	The product of a nuclear reaction inside a burning star

Comprehension

Read the passage about the astronomer, Edwin Hubble, then answer the following questions.

1 How did Edwin Hubble prove the existence of galaxies outside the Milky Way?

2 How did Hubble's observations change the accepted view of the Universe?

3 What other model did Hubble's work help scientists to develop?

4 What is special about the way that the Hubble telescope can be repaired?

The Hubble space telescope was carried into orbit by the space shuttle, Discovery, in 1990. It's named after the American astronomer, Edwin Hubble. Hubble discovered nebula in other galaxies outside the Milky Way, proving the existence of other galaxies and a Universe that was much larger than people believed. Many astronomers at the time opposed this idea. Hubble's observations changed the accepted view of the Universe.

Hubble also made measurements of the expanding Universe. His discoveries have helped scientists to develop the theory of the Big Bang. The Hubble telescope can be serviced by astronauts whilst in its orbit in space. It has been serviced four times and is still working today.

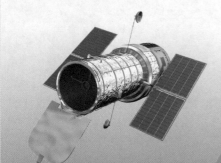

Space and Gravity

Testing Understanding

1 Fill in the missing words to complete the sentences about space and gravity.

a) Inside a star, the force of _____ pulls gases together until the _____ and pressure are so great in the core that _____ reactions occur. Vast amounts of _____ are released as the star changes hydrogen into _____.

b) When a star's _____ has used up all of the available _____ fuel, the star's outer layer cools to a _____ giant and leaves a hot, dense core called a _____ dwarf. Eventually, the white dwarf will also fade away as the star dies.

c) Some stars are massive and instead of becoming a red giant, they _____. This is known as a super nova. If an even larger star explodes, the _____ can be so strong that a black hole is formed and not even _____ can escape!

2 The diagram below shows the path of a comet travelling around the Sun.

a) What can you say about the gravitational pull of the Sun on the comet at A?
b) What can you say about the gravitational pull of the Sun on the comet at B?
c) What can you say about the speed of the comet at A?
d) What can you say about the speed of the comet at B?

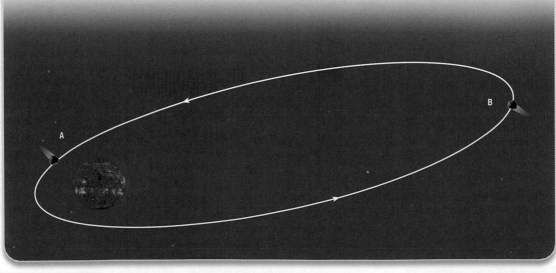

Abby wants to investigate the relationship between the force required to keep a satellite in orbit and its mass.

She plans to do this by setting up a model of a satellite orbiting the Earth. She will use different amounts of Plasticine attached to a string and spin them around. She will use a force sensor to measure the force required.

1. What is the independent variable in this experiment?

2. What is the dependent variable in this experiment?

3. a) Abby thinks that it will be difficult to keep the velocities of each mass she is spinning the same. Explain how she could make sure that the velocities of the masses are constant.

   b) Name one other thing that must be kept constant in order for it to be a fair investigation.

4. What safety precaution should be considered?

5. How could she present her results?

ESSENTIALS

KS3
Science Coursebook Answers
covers all three years

CYCLE OF REPRODUCTION

Page 11 – Quick Test

1. The fusing / joining together of the female nucleus with the male nucleus.

2. External fertilisation takes place outside the bodies of the parent, and internal fertilisation takes place inside the female's body.

3. **Any one from**: It has a tail to enable it to swim; it has structures in its cytoplasm to give it the energy it needs to swim; it has a streamlined shape to help it swim easily; it has a special tip to its head to help it break into the egg cell membrane.

4. To protect the baby from minor bumps.

5. **Any one from**: Ovaries start to release eggs; Breasts start to develop; Hips get broader; Pubic and underarm hair grows

Page 12 – Key Words Exercise

Embryo – The developing egg before it is recognisable as a human baby
Fertilisation – The fusing of the egg nucleus with the sperm nucleus
Foetus – The developing baby once it has limbs and is obviously human
Menstrual cycle – The series of changes that occur in the female reproductive system each month
Ovulation – The release of an egg from an ovary
Placenta – Where substances are exchanged between mothers and foetal blood
Urethra – Carries sperm or urine to the tip of the penis
Uterus – A muscular organ, also known as the womb, where the baby develops

Page 12 – Comprehension

1. One. After fertilisation the egg divides into two separate groups.

2. Two. Each needs to be fertilised by separate sperm.

3. One would develop into a baby the other would not.

4. Two babies would be identical the third would be different.

5. **Any sensible answer, e.g.:** pain; not enough room in the womb; need for Caesarean.

Page 13 – Testing Understanding

1. a) ovaries; ovulation
   b) oviduct; fertilisation / conception
   c) uterus / womb; implant; embryo; foetus
   d) milk
   e) menstrual

2. a) Egg cytoplasm
   b) Sperm tail
   c) Sperm head
   d) Sperm nucleus
   e) Egg nucleus
   f) Jelly layer

Page 14 – Skills Practice

1. **Any one from:** Size and material of container; amounts of food and water; type of shelter; not handled differently.

2.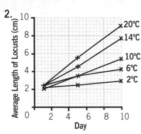

3. The warmer the temperature the faster / bigger / longer they grow

4. Investigate a different range of temperatures (higher) to see if it affects growth.

5. There are five locusts in each so the average is fairly reliable. Mass should also have been measured as length is not necessarily only measure of growth.

ORGANISATION OF LIFE

Page 18 – Quick Test

1. Nucleus, cell membrane and cytoplasm.

2. a) ⊶━━━━━━━━⊛
   b) It is long so that it can transmit messages from place to place.

3. Tissues

4. The moving of pollen grains from the anthers of one flower to the stigma of another.

Page 19 – Key Words Exercise

Cell – The 'building block' of life
Cell membrane – Surrounds the cell and controls what can enter or leave it
Cell wall – A rigid structure made of cellulose around plant cells
Chloroplast – Green structures found in plant cells that absorb light for photosynthesis
Cytoplasm – A jelly-like substance where chemical reactions take place
Nucleus – Controls all the cell reactions and contains the genetic information
Organ – A group of tissues carrying out a particular function
Photosynthesis – The process that green plants use to make sugars (food)
Pollen tube – Grows down the style and carries the male nucleus to the plant ovule
Pollination – The transfer of pollen from an anther to a stigma
System – A group of organs carrying out a particular function
Tissue – A group of similar cells carrying out a particular function

Page 19 – Comprehension

1. Because microscopes were only invented in the 17th Century, and cells can only be seen using microscopes.

2. 1665

3. Because he only saw the dead 'empty' cells.

4. When Shwann and Schlieden suggested that

they are what organisms are made up of.

5. They were thought to have 'just formed' before the theory was accepted.

6. Cells create other cells.

7. **The list should include the following dates and events:**
   1600s Microscope invented
   1665 Hooke uses word 'cell'
   1673 single-celled animal life observed
   1830s Nucleus and cytoplasm observed
   1839 Schwann and Schlieden suggest that all organisms are made of cells
   1855 Virchow says that 'all cells come from cells'
   1890 All life thought to be made up of cells

Page 20 – Testing Understanding

1. a) cell membrane; nucleus; cytoplasm
   b) specialised; tissues; system; organism
   c) nucleus; cytoplasm; daughter
   d) pollen; anther; stigma; pollen tube; style; ovule; fertilisation

2. a) i) nucleus
      ii) cytoplasm
      iii) cell membrane
      iv) cell membrane
      v) cytoplasm
      vi) cell wall
      vii) chloroplast
      viii) vacuole
      ix) nucleus
   b) i) A
      ii) It has no cell wall, chloroplasts or vacuole.

Page 21 – Skills Practice

1. a) Hannah
   b) She used more pollen grains so her results would be more reliable.

2. Number of grains growing tubes.

3. Concentration of sugar.

4. a)

Concentration of Sugar Solution	Number of Tubes
0	0
5	9
10	18
15	14
20	2
25	0

b)

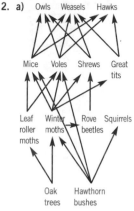

5. The best sugar concentration for germination is 10% as this produced the most tubes.

6. Length of time they were left to grow for; temperature left at; container grown on / in; amount of sugar solution.

INTERACTION IN THE ENVIRONMENT

Page 25 – Quick Test

1. Habitat

2. **Any two from:** thick layer of fat for insulation and food storage; thickly packed feathers to trap air for insulation and to give buoyancy in water; an aerodynamic body (dart-shaped) and webbed feet for swimming; black and white colouring for camouflage.

3. **Any one from:** growing a winter coat; disappearing underground and living off a food store in its bulb; hibernating / sleeping through the winter; shedding leaves; migrating / moving to warmer climates.

4. Green plants that produce food using the Sun's energy.

5. **Any example of a carnivore.**

6. Method of using the relationship between organisms to control pests.

Page 26 – Key Words Exercise

Biological control – Method of using an organism to reduce the number of pests

Carnivore – Animals that eat only animals

Consumer – All animals, because they eat other organisms for energy

Food chain – A way of representing the energy transfer between organisms

Habitat – The particular area in which an organism lives

Herbivore – Animals that eat only plants

Interdependent – Dependent on each other to stay alive in a habitat

Omnivore – Animals that eat both plants and animals

Organic – A process that doesn't use artificial chemicals

Photosynthesis – The process by which plants make their own food using sunlight energy

Predator – An animal that hunts and eats other animals

Prey – An animal that's hunted and eaten by another animal

Producer – A green plant able to produce its own food

Page 26 – Comprehension

1. Prey

2. Predators

3. Herbivores

4. Carnivores

5. The dog brought in a disease which killed most of the wolves.

6. Because the number of moose provide the only food for wolves so when their numbers go up, so do the wolves. As wolf numbers go up, moose numbers go down and so on. It may also depend on the amount of plant food available to the moose.

Page 27 – Testing Understanding

1. **a)** habitat
   b) adapted; temperature; light / water / oxygen / food / nutrients
   c) migrate; hibernation

   d) food; sharp; herbivores; flattened

2. **a)**

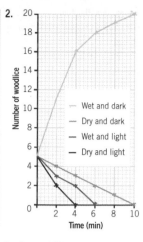

   b) i) **Any one from:** owl; weasel; hawk; mouse; vole; shrew; rove beetle; Great tit.
   ii) They eat other animals.

   c) Mice / voles

   d) Hawthorn ⟶ Winter moth ⟶ Rove beetle ⟶ Shrew ⟶ Owl / Weasel / Hawk

   e) More hawthorn for winter moths, mice and voles, so more of all those organisms higher in the chain, etc.

   f) The Sun

Page 28 – Skills Practice

1. The number of woodlice placed in each condition at the start; the size of the chambers should be equal; the chamber should have been placed on a level surface (so they weren't being tipped into one condition); the students should have timed the 2-minute intervals accurately.

2.

   (Graph: Number of woodlice against Time (min), with lines for Wet and dark, Dry and dark, Wet and light, Dry and light)

3. Dry and light.

4. Woodlice prefer wet and dark conditions.

5. **a)** No
   b) Because it wouldn't have been so reliable.

SORTING DIFFERENCES

Page 32 – Quick Test

1. Because they can't reproduce together to form fertile offspring. / Because they produce infertile offspring when they reproduce together.

2. Genetic / inherited factors, and environmental factors.

3. Created a system of classifying organisms.

4. **Any two from:** hair-covered bodies; warm blooded; give birth to live young; suckle their offspring.

Page 33 – Key Words Exercise

Amphibian – An animal with moist skin and lungs; it returns to water to breed

Bird – An animal with feathers, wings and a beak

Classification – Grouping living things using features they have in common

Fish – An animal with fins and gills

Infertile – An organism unable to produce young

Invertebrate – An animal that doesn't have a backbone

Kingdom – The first grouping (of five) in the classification of living things
Mammal – A hair-covered animal that suckles its young
Multi-cellular – A body made up of many cells working together
Reptile – An animal with dry, scaly skin and lungs
Species – A group of similar organisms able to produce fertile young
Variation – Differences between individuals of the same species
Vertebrate – An animal that has a backbone

Page 33 – Comprehension

1. To make identification of all living things easier to carry out using a key.

2. Because it is only relative and there may be a lot of variation in one group.

3. *Homo*

4. More alike / similar.

5. To identify plants and animals

Page 34 – Testing Understanding

1. a) similar; fertile
   b) variation; genetic / inherited; hair / eye / skin; environmental
   c) characteristics / features; classification; kingdoms
   d) hair; live; suckle

2. a) Yes, there is a positive correlation.
   b) i) Point F
      ii) Point A
   c) i)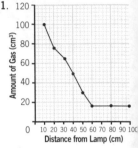
      ii) They most fit the pattern.
      iii) They least fit the pattern.

Page 35 – Skills Practice

1. Millimetres

2.

Length of Little Finger (mm)

3. a) The first group was bigger than the next; The last two groups overlapped.
   b) 51–55; 56–60; 61–65; 66–70; 71–75

4. So that the measuring method was the same each time (fair test)

5. The measurements are spread out fairly evenly amongst the class but there is quite a large range of little finger lengths in the class.

6. Genetic. This is due to body size, etc. and not likely to be affected by the environment at all.

FOOD AND LIFE

Page 39 – Quick Test

1. **Any three from:** movement; respiration; sensitivity; growth; reproduction; excretion; nutrition.

2. Brain, spinal cord and nerves.

3. Fibre (roughage).

4. Chlorophyll.

Page 40 – Key Words Exercise

Amino acids – The small molecules produced after protein digestion
Balanced diet – Contains the right proportions of the seven food types for you
Carbon dioxide – A gas that plants need in order to photosynthesise
Chlorophyll – The green pigment that absorbs light energy
Digestion – The breakdown of large food molecules into small molecules so they can be absorbed
Enzymes – Chemicals that break

---

down large, insoluble molecules into small, soluble molecules
Fibre – Helps push food through the intestine and prevents bowel problems
Nitrogen – An element necessary for plants to make protein
Photosynthesis – The process by which plants make food
Protein – A substance in the diet used for growth and repair of body tissues

Page 40 – Comprehension

1. The (small) intestine.

2. Sugar wasn't put into the tubing; the starch hadn't been broken down.

3. The blood(stream).

4. The amylase (enzyme) broke down the starch into sugar.

5. Sugar molecules are small enough to get through the small pores (holes) in the tubing but starch molecules are too large to get through.

6. It was completely broken down into sugar.

7. Quite well. But lots more happens in the gut, and the blood is contained in vessels (capillaries) and is constantly moving.

Page 41 – Testing Understanding

1. a) balanced; seven
   b) carbohydrates; proteins; vitamins
   c) overweight / obese; water
   d) molecules; enzymes; absorbed; blood(stream)
   e) photosynthesis; carbon dioxide; water; (sun)light; oxygen

2. a)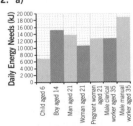

---

   b) **Any sensible answer, e.g.:** He could have a larger body / manual job / more active lifestyle.
   c) The clerical worker's job is less physically active than the manual worker's job, so he needs less energy.
   d) She has to supply the energy needs of the baby in the womb, as well as her own.

Page 42 – Skills Practice

1.

Distance from Lamp (cm)

2. Oxygen

3. As light intensity decreases, less gas is collected, so the rate of photosynthesis decreases / As light intensity increases, more gas is collected, so the rate of photosynthesis increases.

4. a) $70cm^3$
   b) $15cm^3$

5. The amount of gas collected.

6. The light bulb made no difference – extra light from the surroundings was causing the photosynthesis.

7. The temperature of the water; the direction of the light; the amount of water; the carbon dioxide levels in water; the surrounding light level / temperature; the air pressure.

RESPONDING TO THE ENVIRONMENT

Page 46 – Quick Test

1. A response.

2. Behaviour

3. When they're contracting / getting smaller.

4. Antagonistic pairs.

5. Oxygen

Page 47 – Key Words Exercise

Antagonistic pair – Two muscles that work opposite to each other
Behaviour – A pattern of actions carried out by an animal
Carbon dioxide – A gas that's a waste product of respiration
Excretion – Removal of waste products of chemical reactions from the body
Glucose – The energy-rich molecule used in respiration
Innate behaviour – Behaviour that's instant and caused by genes
Learned behaviour – Behaviour that occurs as a result of experience
Motor nerve – A type of nerve that causes an action to take place
Muscle contraction – When a muscle shortens to exert a force
Oxygen – A gas used up to help break down glucose in respiration
Receptor – The end of a sensory nerve that transforms one type of energy into a nerve impulse
Respiration – A combination of chemical reactions that release energy in a cell
Response – An action as a result of a nervous impulse, for example, moving a muscle
Stimulus – Something that causes a response in the nerve cell

Page 47 – Comprehension

1. It was a response to the sight and smell of the food; they were ready to begin digestion.

2. The presence of food and the sound of the bell.

3. The original dogs already showed this behaviour, so he had to use other dogs to show that they could learn to associate food with the sound of a bell.

4. It would make them produce saliva in readiness for eating.

5. **Any sensible answer, e.g.:** The sound of cooking in the kitchen might make you salivate.

Page 48 – Testing Understanding

1. a) response; stimulus; behaviour
   b) nerves; receptor; impulse
   c) contracts; relaxes; pull; antagonistic
   d) respiration; Oxygen; glucose; fatigue / cramp
   e) innate; learned

2. a) and b) See diagram below.
   c) **Any sensible answer, e.g.:** A hot object, like a plate, which would cause the hand to move away quickly; a sharp object, like a pin, which would also cause the hand to move away quickly.
   d)

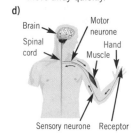

Brain — Motor neurone
Spinal cord — Hand
Muscle
Sensory neurone — Receptor

Page 49 – Skills Practice

1.

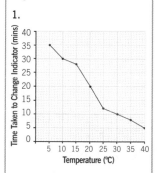

Time Taken to Change Indicator (mins) vs Temperature (°C)

2. Carbon dioxide is given out more quickly as the temperature increases.

3. Woodlice are more active at higher temperatures, so they respire more quickly as they use up more energy.

4. Temperature.

5. It would harm the woodlice.

6. The number of woodlice; the amount of the light; the amount of indicator; the position of the gauze; the size of the tube; the bung must be inserted tightly to prevent air getting in or out.

SURVIVAL IN THE ENVIRONMENT

Page 53 – Quick Test

1. Variation.

2. Genetics / genes / characteristics passed on from parents.

3. Chromosomes.

4. The environment / surroundings / lifestyle.

Page 54 – Key Words Exercise

Antibiotics – Drugs, produced from microbes, that are used to control bacterial infections
Characteristics – Those features that can be seen in or on an organism
Chromosome – A thread-like structure, like a string of beads
Environmental factors – Features of the surroundings that may affect an organism
Genes – Units that determine an organism's characteristics
Genetic factors – Influences on the genetic material of an organism
Inheritance – The passing on of characteristics from one generation to the next
MRSA – A disease often found in hospitals, caused by a bacterium
Nucleus – The part of the cell where the genetic material is found
Resistance – The ability to be relatively unaffected by something, for example, a drug
Variant – An organism that shows a certain difference in a species
Variation – Differences within a single species

Page 54 – Comprehension

1. His brown hair; green eyes; straight nose; large ears.

2. His blond highlights; pink skin; pierced eyebrow; scar on knee.

3. His height; skinny build; his ability to run fast; basketball ability.

Page 55 – Testing Understanding

1. a) characteristics / features
   b) genes; genetic
   c) environmental
   d) species; survive
   e) nature; nurture

2. a) **Any sensible answer, e.g.:** A new predator killed them – predators aren't likely to have killed all dinosaurs because then they would have no food (predator-prey relationship). In addition, there was a huge variety of dinosaurs, so there would have had to have been a variety of predators; a new disease killed them – due to the variety of dinosaurs, a single disease is unlikely to have affected them all; new animals ate all their food – the diet of dinosaurs was likely to be varied, so it's unlikely that new animals would eat the food of all the different dinosaurs.
   b) The Sun being blocked out could have reduced the temperature of the Earth to a level that killed the dinosaurs. It would also affect some plants because there wouldn't be sunlight for photosynthesis, although it's possible some could survive (and the evidence is that they did). As food became scarce, the dinosaurs would starve.

Page 56 – Skills Practice

1. The camouflaged 'insects' seemed to survive best as the numbers remaining on the same coloured cards were generally higher. The birds seemed to prefer the red and blue 'insects' as the yellow 'insects' appeared to be eaten less on other backgrounds.

2. To make sure it was only birds eating the 'insects' and they weren't disappearing for other reasons.

3. By repeating the exercise or by using more 'insects'.

4. That weather conditions (e.g. shade and humidity) were the same for all 'insects'; that the size and shape of the 'insects' and the cards were the same; that the colours were the same; that the placing of the 'insects' was completely random; that the 'insects' were the same distances away from plants and bushes.

ENVIRONMENTAL RELATIONSHIPS

Page 60 – Quick Test

1. A particular area where an organism lives.

2. The total number of individuals of one species living in a particular area.

3. Quadrats.

4. Some is lost due to respiration / heat; some is lost in kinetic (movement) energy / urine and faeces.

5. The mass of individuals of a particular type in a food chain.

Page 61 – Key Words Exercise

Biodiversity – The range of different species in an area
Community – The total number of populations in a habitat
Distribution – The spread of individuals of a species in an area
Ecosystem – The community of organisms together with the physical conditions in a defined area
Food chain – A sequence of organisms showing 'what eats what' in an area
Habitat – The particular type of area in which an organism lives
Immigration – To move into an area
Population – The total number of individuals of the same species in a habitat
Primary consumers – A name given to animals that eat producers

Producers – A name given to green plants that can make their own food
Pyramid of biomass – A way of showing the relative mass of organisms in a food chain or web
Pyramid of numbers – A way of showing the relative numbers of organisms in a food chain or web
Quadrat – A device for sampling 'fixed' organisms in a large area
Secondary consumers – A name given to animals that eat herbivores

Page 61 – Comprehension

1. Photosynthesis.

2. Conditions in the rainforests are better for photosynthesis; the forests get more sunlight and rain (water); the temperature is more stable; nutrients decompose more quickly for uptake by plants.

3. Herbivores (primary consumers).

4. Carnivores (secondary consumers).

5. As heat in respiration; in kinetic (movement) energy; in excretion (urine and faeces).

Page 62 – Testing Understanding

1. a) adapt; survival
   b) community
   c) distribution; habitat
   d) sample; quadrat
   e) numbers; biomass; energy
   f) ecosystem; thermometer; light meter
   g) recapture

2. a)
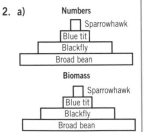

b)
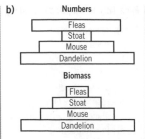

Page 63 – Skills Practice

1.
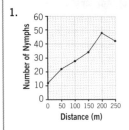

2. Temperature; pH; flow rate; food supply.

3. The sample distance.

4. Yes. The numbers did increase as they sampled away from the town.

5. There was more pollution in that area; more predators in that area; it was purely a random difference (i.e. no particular reason).

6. The town might produce pollution that would reduce the oxygen levels in the water.

DISRUPTION OF LIFE PROCESSES

Page 67 – Quick Test

1. A disease-causing microbe

2. Bacteria, viruses and fungi

3. Nicotine

4. Alveoli

5. Liver and brain

6. Antibodies

Page 68 – Key Words Exercise

Alveoli – The air sacs where oxygen enters and carbon dioxide leaves the blood
Antibodies – Substances produced by white blood cells that can neutralise a specific microbe

Antitoxins – Chemicals that neutralise a poison produced by a microbe
Bronchitis – A condition in which air sacs are damaged due to constant coughing
Emphysema – A smoking-related disease caused by excessive mucus and smoke particles
Hallucinogens – Substances that cause hallucinations
Immunity – Protection from disease by the production of antibodies
Nicotine – An addictive chemical found in cigarette smoke
Pathogen – A disease-causing microbe
Vaccine – A collection of dead or weakened microbes, or antibodies, injected to give protection

Page 68 – Comprehension

1. **Any sensible answer, e.g.:** by using smoking by well-known celebrities, or in places that are well known, to make it appear a 'good' thing.

2. The groups should have had equal numbers of people; the people should have been of the same ages; the people should have had similar builds / lifestyles; there should have been the same numbers of males and females.

3. The study showed that smoking led to an increase in the likelihood of contracting lung cancer, so it could have put people off buying cigarettes.

4. Not true – it only increases the chances of getting lung cancer.

Page 69 – Testing Understanding

1. a) **In any order:** bacteria; viruses; fungi
   b) bacteria; viruses; fungi
   c) white blood; antitoxins; antibodies; stick / attach; digest
   d) immunisation / vaccination; dead / killed; antibodies
   e) alcohol; drugs; addicted

2. a)

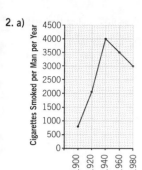

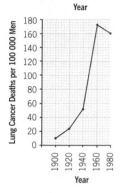

b) The number of cigarettes smoked per year rose steeply until 1940 and then began to drop.

c) The number of deaths rose until 1960 and then began to drop.

Page 70 – Skills Practice

1. **Any one from:** they could measure the diameter of each area with a ruler; they could measure the area of each by putting it over graph paper and counting the squares.

2. C as it had the biggest clear area.

3. As a control to show that it was the penicillin, not the liquid-soaked disc, that caused the effect.

4. They were using bacteria that could have harmed them.

5. To kill any harmful bacteria on their hands / to prevent contamination from their hands.

VARIETY IN THE ENVIRONMENT

Page 74 – Quick Test

1. Sex cells

2. Germination

3. An individual that is genetically identical to another.

4. A chemical that kills only certain plants.

5. A chemical that kills pests.

6. The use of one organism to control the numbers of another.

Page 75 – Key Words Exercise

Asexual reproduction – Production of new individuals requiring only one parent
Biological control – Using a living organism to keep pest numbers down in a crop
Cloning – Producing exact genetic copies of an organism
Crop yield – The amount of crop produced in a growing season
Cuttings – Small sections of a plant grown into separate copies in different places
Gametes – Reproductive cells that fuse together at fertilisation
Herbicides – Substances used to kill plants
Minerals – Substances that are necessary for chemical reactions such as photosynthesis
Pesticides – Substances used to kill animals such as insects, slugs or other pests
Selective – Something that only affects certain types of plant or animal, not all of them
Sexual reproduction – Production of new individuals requiring male and female parents
Territory – An area needed by animals to allow feeding and breeding

Page 75 – Comprehension

1. To prevent weeds from growing and competing with the strawberries.

2. **a)** Water and minerals / nutrients

b) Fewer resources would get to the flowers / fruit, so there would be less fruit.

3. They're genetically identical to the parent.

4. Biological control

Page 76 – Testing Understanding

1. **a)** gametes; sperm / eggs; eggs / sperm; moving / movement; food store
   b) fertilisation; nuclei; unique; genes / chromosomes / DNA; parents; characteristics / traits
   c) parent / individual / organism; animals
   d) identical; differences; cuttings

2. **a)** A weedkiller that only kills certain plants / types of plant.
   b) **Any one from:** weed leaves are broader than grass leaves; grass leaves are narrower than weed leaves.
   c) To improve the appearance of the lawn; to remove plants that are competing with the grass / to keep only grass plants in the lawn.
   d) The grass could disappear as buttercups would outcompete it for resources, especially light.
   e) It's harmful / poisonous to us.

Page 77 – Skills Practice

1.

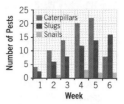

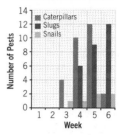

2. Temperature; soil pH; the amount of light; the amount of nutrient in the soil

3. The number of pests

4. Yes. The numbers of pests were lower in each case, although there was little difference in snail numbers.

5. Yes. The amounts were quite different provided that all other factors were the same (for example, none of the lettuces in the water-sprayed plot were eaten by other animals / birds, etc).

6. The lettuces got bigger, so there was more food available.

7. **Any one from:** they got eaten by birds / other animals; they were killed by a disease; they hatched into butterflies after week 5.

MANIPULATING THE ENVIRONMENT

Page 81 – Quick Test

1. Breeding for a desired characteristic.

2. The loss of a gene from a population.

3. Carbon dioxide

4. Oxygen

5. Because so much energy is lost at each step.

6. A genetically modified organism

Page 82 – Key Words Exercise

Bioaccumulation – An increase in the concentration of something along a food chain

Embryonic – Originating from the developing fertilised egg

Eutrophication – Caused by fertilisers in water, leading to stagnation

Global warming – An increase in the Earth's temperature caused by the thickening of the layer of carbon dioxide in the atmosphere

GMO – An organism that has had its genetic make-up artificially altered

Limiting factors – Environmental changes that alter a reaction like photosynthesis

Selective breeding – Using variation to breed a desired characteristic into the offspring

Stem cell – A cell able to develop into many different types

Sustainable development – Meeting human needs without damaging the environment

Varieties – A range of different types within the same species

Page 82 – Comprehension

1. Able to be broken down / decayed / decomposed in the soil.

2. To allow the microorganisms of decay / decomposition to respire.

3. The compost contains the broken down products from the decomposed plants.

4. They would slow / stop photosynthesis.

5. Organic cabbages are likely to be a little more 'eaten' in appearance / have holes in the leaves / be different sizes rather than a uniform size.

Page 83 – Testing Understanding

1. a) yield
   b) pests; vegetables; greenfly / blackfly
   c) food; insects; food; reduced / lessened; slugs;

fewer; more

d) Pesticides; diluted; concentrated; kill; bioaccumulation

2. a)

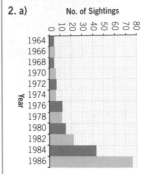

   b) **Any one from:** 59; 60 (**not 59.5**).
   c) 1983
   d) 12
   e) The small mammals ate insects, so the DDT accumulated in their bodies. The peregrines then ate lots of mammals (bioaccumulation).

Page 84 – Skills Practice

1.

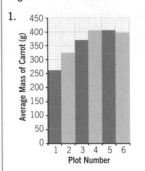

2. Up to 12g made a difference but this was probably the most that the carrots could take up from the soil, so the extra had no effect.

3. It could wash out into streams and rivers and cause pollution / eutrophication.

4. The mass of the carrots

5. The size of the plots; the amount of water / light; the number of seeds / plants planted; the amount of nutrient in the plot before the experiment; the amount of weeding.

SOCIAL INTERACTION

Page 88 – Quick Test

1. Behaviour

2. A stimulus

3. It stays inside its food source, away from predators

4. Courtship

5. Social (behaviour)

Page 89 – Key Words Exercise

Behaviour – The pattern of responses of individuals to a stimulus or stimuli

Courtship display – Behaviour designed to allow reproduction to occur

Fertilisation – The fusing of male and female nuclei

Habit – Learned behaviour gained by a series of steps

Mating – The meeting of a male and female for reproduction

Population – The total number of individuals of the same species in an area

Response – The action made due to receiving a stimulus

Species – Similar organisms able to breed to produce fertile offspring

Status – An individual's position within a group

Stimulus – Something that causes a response in the nervous system

Page 89 – Comprehension

1. The period when male and female deers are found together for mating.

2. a) They are able to see if they are well matched and one can back down without a fight if necessary, so preventing injury.
   b) The strongest stag mates with more deer. The 'best' genes are then passed on to the next generation.

3. To attract females (hinds) and to maintain the group.

4. To feed it with milk and so that it can learn the necessary behaviour for survival.

Page 90 – Testing Understanding

1. a) behaviour; changes
   b) responses; protect / guard
   c) learn / adopt; status; mating
   d) social

2. a)

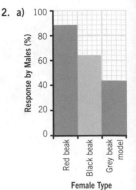

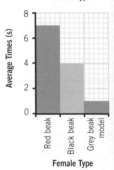

   b) The red-beaked female caused more response than the black-beaked female and the males didn't pay much attention to the model.
   c) The length of the courtship behaviour followed the same pattern as the number of responses (i.e. the courtship behaviour of males towards the red-beaked females lasted longer than towards the black-beaked females, and the shortest courtship behaviour was towards the grey-beaked model).
   d) The male finches respond to red-beaked females more than to black beaked and aren't responsive to models or grey-beaked females.
   e) The female finches and the colours of their beaks.

Page 91 – Skills Practice

1.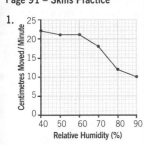

2. Temperature; amount of light; amount of liquid in the dish; the woodlice were of a similar size.

3. Centimetres moved per minute.

4. They should have been kept in identical conditions, preferably at a relative humidity of less than 40%.

5. a) To remove any interaction between individuals in the dish.

   b) To make sure the results were reliable; to make sure the results were all the same pattern; to remove the effect of any 'freak' results (**to make it a fair test' is not acceptable**).

6. They move less in conditions of higher humidity, which would help them retain water for gas exchange / breathing. In addition, the higher the humidity, the more likely it is that the place would be dark and sheltered, so they would be less likely to be eaten by predators.

ACIDS AND ALKALIS

Page 96 – Quick Test

1. Alkali

2. To warn people of the dangers of using them.

3. Water / alcohol.

4. The strength of an acid or an alkali.

Page 97 – Key Words Exercise

Acid – A chemical that has a pH of less than 7

Alkali – A chemical that has a pH of more than 7

Corrosive – A substance that attacks and wears away living materials, metals and rocks

Dilute – To water down

Hazard symbol – A symbol used to warn people about the chemicals in a container

Hazardous – Something that might cause an accident to happen

Indicator – A chemicals that's one colour when mixed with an acid and another colour when mixed with an alkali

Lime – An alkaline compound used to neutralise acidic soils

Neutral – A chemical that has a pH of 7 and is neither acidic nor alkaline

Neutralise – To cancel out

pH – A scale used to measure how acidic or alkaline a solution is

Sour – A sharp, acidic taste

Universal indicator – A mixture of indicators that can be used to test how acidic or alkaline a solution is

Page 97 – Comprehension

1. Roadsides and areas of rough ground.

2. Nitrogen-rich soils.

3. By touching a nettle leaf and breaking the delicate hairs that cover it.

4. They contain an acid.

5. They are large, dark green and have an oval shape.

6. By rubbing it with a crushed dock weed leaf.

7. They contain an alkali which neutralises the acid of the nettle skin.

Page 98 – Testing Understanding

1. a) alkalis
   b) less; weak; sour; corrosive / hazardous
   c) more / greater; weak
   d) 7
   e) alkaline; indicator; alkaline
   f) red; neutral; purple

2. a)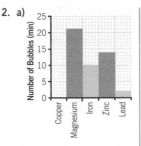

   b) magnesium, zinc, iron, lead, copper

Page 99 – Skills Practice

1.

Antacid	pH
A	5
B	7
C	6
D	4

2.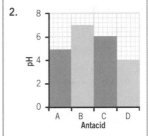

3. The type of antacid

4. The pH

5. Same volume of acid / same type of acid / same pH meter.

6. Measuring cylinder

7. B

CHEMICAL REACTIONS

Page 103 – Quick Test

1. Irreversible

2. It turns cloudy.

3. magnesium oxide

4. Fuel, oxygen, heat.

Page 104 – Key Words Exercise

Chemical reaction – A change that makes new substances

Condense – A change of state from gas to liquid

Corrosive – A substance that attacks and wears away living materials, metals and rocks

Fuels – Substance that can be burned to release energy

Irreversible – A change that can't be undone

Limewater – A chemical used to test for the presence of carbon dioxide

Melting – A change of state from solid to liquid

Oxide – Compound made when a substance reacts with oxygen

Products – The chemicals at the end of a reaction

Reactants – The chemicals at the start of a reaction

Reversible – A change that can be undone

Page 104 – Comprehension

1. Bonfire night, Diwali and New Years' Eve.

2. A substance that releases energy when it's burned.

3. a) A chemical that releases oxygen.
   b) They allow the fuel to burn better.

4. When they are heated they can release bursts of coloured light.

5. They are explosives.

Page 105 – Testing Understanding

1. a) reactants; products
   b) reversible
   c) oxygen; oxide; copper
   d) carbon
   e) oxygen
   f) sulfur

2. a)

Metal	Start Temp. (°C)	End Temp. (°C)	Temp. Change (°C)
Zinc	20	35	15
Iron	20	28	8
Magnesium	21	47	26
Copper	20	20	0

b)

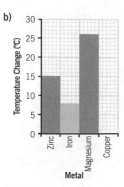

c) Copper

Page 106 – Skills Practice

1.

Volume of Water (cm³)	Temperature of Water after 2 Minutes (°C)
25	75
50	60
75	45
100	30

2. The volume of water.

3. The temperature of the water after 2 minutes.

4. Heating all the beakers for the same length of time / using the same starting temperature.

5. Thermometer

6. The less water there is the higher the temperature reached / the more water there is the lower the temperature reached.

PARTICLES

Page 110 – Quick Test

1. Liquid

2. Diffusion

3. It increases.

Page 111 – Key Words Exercise

Attraction – Force that holds matter together
Compressible – Can be squashed
Contract – Get smaller
Density – Equal to mass divided by volume
Diffuse – To mix thoroughly
Evaporate – A change of state from liquid to gas
Expand – Get bigger

Incompressible – Can't be squashed
Particle theory – The model used to explain why solids, liquids and gases behave as they do
Pressure – Equal to force divided by area
Vibrate – To move up and down

Page 111 – Comprehension

1. Hg

2. From its Latin name *hydragyrum*.

3. It is a liquid at room temperature.

4. By heating the mineral cinnabar.

5. Because it is a liquid it expands and contracts more when heated or cooled.

6. Because of health concerns.

7. To make amalgams for fillings.

8. They thought they could turn it into gold.

9. Mercury was used in the hat-making industry and one of the symptoms of mercury poisoning is madness.

Page 112 – Testing Understanding

1. **a)** particles
   b) volume; close; incompressible; vibrate
   c) shape; flow
   d) gas; particles; compressible

2. **a)**

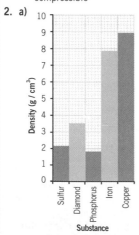

   b) i) Iron and copper
   ii) They are both metals

Page 113 – Skills Practice

1.

Gas in the Balloon	Distance the Balloon Travels (m)
Helium	+1.5
Neon	-0.6
Argon	-0.8
Krypton	-1.3

2.

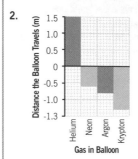

3. Gas used

4. The distance the balloon travels

5. Same volume of gas / same type of balloon / same time

6. Timer / stopwatch / stop clock

7. Krypton

SOLUTIONS

Page 117 – Quick Test

1. False

2. Insoluble

3. Distillation

4. To separate the inks in a dye / detect illegal drugs or forgeries.

5. **Any one from:** increase the temperature of the solvent; stir the mixture; use smaller pieces of solute; increase the surface area.

Page 118 – Key Words Exercise

Chromatography – A technique used to separate different coloured inks
Condense – A change of state from gas to liquid
Dissolve – To mix a substance with a solvent to make a solution
Distillation – A process used to separate a solvent from a solution
Evaporate – A change of state from liquid to gas

Filtering – Used to separate an insoluble solid from a solution
Saturated solution – A solution that can't hold any more solute at a particular temperature
Soluble – A term used to describe substances that dissolve
Solute – A solid that dissolves to make a solution
Solution – Formed when a solute dissolves in a solvent
Solvent – A liquid that dissolves a solute to make a solution

Page 118 – Comprehension

1. They could be damaged by water.

2. No water is involved.

3. **a)** Perchloroethylene
   b) Because it says that the solvent is good at dissolving greasy and oily stains.
   c) Health / environmental concerns; irritates some people's eyes and throats.

Page 119 – Testing Understanding

1. **a)** particles
   b) fixed; separate; gases; granite
   c) solution; solute; solvent
   d) solvent; insoluble; solvents

2. **a)**

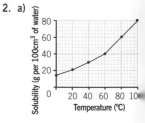

   b) As the temperature increases the solubility of copper sulfate increases.

Page 120 – Skills Practice

1.

Temperature (°C)	Time for Salt to Dissolve (s)
20	18
35	13
50	8

2.

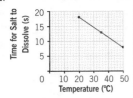

3. Temperature

4. Time for the salt to dissolve.

5. **Any one from**: the volume of water; the mass of salt.

6. Measuring cylinder.

7. The higher the temperature of the water the less time it takes for the salt to dissolve.

ATOMS AND ELEMENTS

Page 124 – Quick Test

1. It's made of only one type of atom.

2. In order of increasing atomic number.

3. It's a metal.

4. a) Carbon
   b) Oxygen
   c) Calcium
   d) Iron

5. Magnesium oxide.

Page 125 – Key Words Exercise

Alloy – A mixture of metals
Atom – A very small particle
Atomic number – The number of protons in an atom
Compound – Contains atoms of two or more different elements that have been joined together by a chemical reaction
Element – A material that's made of only one type of atom
Molecule – A small group of atoms that are joined together
Oxide – Made when a material is burned in air
Periodic table – A way of displaying elements in order of increasing atomic number
Product – The substance made by a chemical reaction
Reactant – The chemical used up during a reaction
Symbol – A one or two-letter code used to represent an element

Page 125 – Comprehension

1. Siberia

2. St. Petersburg

3. He designed the first version of the periodic table.

4. 63

5. For elements that hadn't been discovered. He made predictions about their properties.

6. They hadn't been discovered.

7. Element 101 (mendelevium).

8. A crater on the Moon.

Page 126 – Testing Understanding

1. a) atoms; elements
   b) periodic; non; left
   c) shiny; mercury
   d) electricity; oxygen; oxide

2. a) A and C   b) C
   c) A   d) B and D

Page 127 – Skills Practice

1. magnesium oxide.

2. a) and b)

Mass of Magnesium Burned (g)	Mass of the Magnesium After it's Burned (g)	Difference in Mass (g)
0.6	1.0	0.4
1.2	2.0	0.8
1.8	3.5	1.7
2.4	4.0	1.6

3. a) and b)

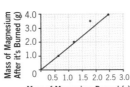

c) The more magnesium that's burned, the higher the mass of magnesium oxide that's made.

COMPOUNDS AND MIXTURES

Page 131 – Quick Test

1. Magnesium oxide.

2. Sodium chloride.

3. It has one atom of carbon and two atoms of oxygen.

4. By the fractional distillation of liquid air.

Page 132 – Key Words Exercise

Atom – A very small particle
Chemical reaction – A change in which new substances are made
Compound – Contains atoms of two or more different elements that are chemically joined
Condense – Turn from a gas to a liquid
Element – A substance made of only one type of atom
Formula – A code to represent the type and number of atoms present
Fractional distillation – A way to separate mixtures of liquids that have different boiling points
Magnetic – A word describing a material that's attracted by a magnet
Mixture – Two or more different substances that are mixed together but not chemically joined
Precipitate – An insoluble solid made when two solutions react together

Page 132 – Comprehension

1. By the fractional distillation of liquid air.

2. Each molecule contains one nitrogen and three hydrogen atoms.

3. A chemical that increases the rate of reaction but isn't used up itself.

4. To help plants grow better; to increase the crop yields; to replace nutrients that have been used up as the plants grow.

5. To make them grow better; to help them produce lots of green leaves; to increase crop yields.

6. It's stunted; it doesn't grow well; it goes yellow.

Page 133 – Testing Understanding

1. a) Elements; atoms; reaction
   b) oxygen; carbon
   c) composition; hard
   d) mixture; mixtures
   e) composition; easy
   f) nitrogen; distillation

2. a) C
   b) A
   c) B
   d) D

Page 134 – Skills Practice

1.

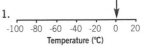

2. a) A thermometer.
   b) A measuring cylinder.
   c) A balance / scales.

3.

	Factor to Change	Factor to Keep Same	Factor to be Measured
Temp. the Water Boils at			✓
Volume of Water		✓	
Mass of Salt	✓		

4. a) and b)

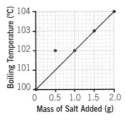

5. To check his results.

ROCKS AND WEATHERING

Page 138 – Quick Test

1. Igneous and metamorphic.

2. You see air bubbles and the mass of the rock increases.

3. Water gets into cracks. When it freezes, it expands and further forces open the cracks. This is freeze-thaw action.

4. When sulfur dioxide or nitrogen oxides dissolve in rainwater.

5. A slow-moving river of ice.

6. They get smaller / smoother / more rounded.

Page 139 – Key Words Exercise

Deposit – Lay down
Evaporates – Minerals formed when water that contains dissolved salts evaporates
Fossil fuels – Energy resources including coal, oil and natural gas
Fossils – The remains of dead plants and animals
Freeze – Turn from liquid to solid
Glacier – A slow-moving river of ice
Interlocking – Joined together with no gaps
Mineral – A naturally-occurring solid compound with a crystalline structure
Porous – Contains gaps between grains that air or water can enter
Sediment – Solids that are deposited by a river

Page 139 – Comprehension

1. Lyme Regis, Dorset.

2. To sell them to earn money for their family.

3. Because the cliffs were unstable and could collapse, burying the children.

4. It was a crocodile-like dinosaur.

5. A plesiosaur and a pterosaur.

6. Species existed in the past that don't exist today.

Page 140 – Testing Understanding

1. a) rocks; sedimentary; together
   b) metamorphic; igneous; crystals; gaps / spaces
   c) cracks; solid; expands; further / more / wider; onion
   d) dioxide; lowers / reduces; carbonates
   e) acid; pH

2. a) (Shelly) limestone because it's at the bottom / the other rocks are on top of it.
   b) The remains of dead plants and animals that lived a long time ago and were preserved in rock.
   c) It was formed in the sea / marine conditions.
   d) They were made at the same time / are the same age.
   e) Water can get into existing cracks. If the temperature drops, the water can freeze to form ice. As the ice expands, it forces the existing cracks even further apart.

Page 141 – Skills Practice

1.
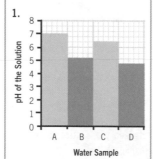

pH of the Solution — Water Sample A, B, C, D

2. It's more accurate / more precise. It logs the results, so you don't have to record them.

3. oxygen; carbon dioxide

THE ROCK CYCLE

Page 145 – Quick Test

1. Sedimentary.

2. High temperatures and high pressures.

3. Slate.

4. Magma.

5. Lava.

6. It cooled slowly below ground.

Page 146 – Key Words Exercise

Cement – 'Glue' that sticks sediments together
Deposit – Lay down
Evaporates – Minerals formed when water that contains dissolved salts evaporates
Fossils – The remains of dead plants and animals
Lava – Molten rock above the Earth's surface
Magma – Molten rock below the Earth's surface
Marble – Type of metamorphic rock formed from limestone
Sedimentary – Type of rock formed when sediments are cemented together by dissolved salts
Slate – Type of metamorphic rock formed from mudstone

Page 146 – Comprehension

1. The Pacific.

2. Volcanoes and earthquakes.

3. The core.

4. A few centimetres per year (the same rate as your fingernails grow).

5. The plates try to move past each other but get stuck. The forces build up until the plates suddenly move, releasing the energy that has built up as an earthquake.

6. The plates move towards each other and one plate is forced beneath the other.

Page 147 – Testing Understanding

1. a) rocks; pressure; salts
   b) crumbly
   c) pressures; limestone; mudstone
   d) crystals; destroyed
   e) cools; magma; large / big
   f) lava; small

2. a) Sedimentary – Grains of sediment are cemented together by dissolved salts
   Metamorphic – Existing rocks are changed by high temperatures and pressures
   Igneous – Molten rock cools down and solidifies
   b) The remains of a dead plant and animal that is preserved in rocks.
   c) Sedimentary.
   d) Igneous. The molten rock would destroy any organisms.

Page 148 – Skills Practice

1. Ruler / tape measure.

2.

Crystal Size (mm) — Rock Sample A, B, C, D

3. Sample B. It has the largest crystals so it cooled most slowly because the crystals had time to grow.

4.

Type of Rock that was Used	Rock Sample	Crystal Size	Factor that was Changed	Factor that was Measured	Factor that was Kept the Same
✓ (all are igneous rocks)			✓	✓	

METALS AND METAL COMPOUNDS

Page 152 – Quick Test

1. Hydrogen

2. Zinc sulfate

3. Sodium chloride

4. Copper nitrate

5. Alkali + acid ➝ salt + water

Page 153 – Key Words Exercise

Acid – The chemical opposite of bases; these compounds all contain the element hydrogen
Alkali – A soluble base
Carbon dioxide – A gas, produced when a metal carbonate reacts with an acid, which turns limewater cloudy
Element – A substance that's made of only one type of atom
Evaporates – Turns from a liquid to a gas

Formula – A way of representing the type and number of atoms in the smallest particle of a substance

Hydrogen – A gas, produced when a metal reacts with an acid, which burns with a squeaky pop

Insoluble – Doesn't dissolve

Metal carbonate – A chemical compound that contains metal, carbon and oxygen atoms

Symbol – A one or two-letter code used to represent an element

Page 153 – Comprehension

1. Sodium chloride
2. Breakfast cereals; ready-made sandwiches; pasta sauces
3. 6g
4. Less than 6g
5. 50 000
6. Stroke; heart disease; eye damage; kidney damage

Page 154 – Testing Understanding

1. a) good; graphite
    b) type; two; one; four
    c) hydrogen; squeaky; nitrate
    d) salt; limewater
    e) water; chloride; gas

2. a) iron chloride; hydrogen
    b) Bubble it through limewater and it turns the limewater cloudy.
    c)

    Chromium (2 segments)   Nickel (1 segment)   Iron (7 segments)

Page 155 – Skills Practice

1.

Test Tube	Any Water?	Any Oxygen?
1	No	Yes
2	Yes	No
3	Yes	Yes

2. The iron nail only rusts when water and oxygen are present.

3. There would be little or no rust because the layer of oil stops the water and oxygen from reaching the iron nail.

THE REACTIVITY OF METALS

Page 159 – Quick Test

1. Any sensible answer, e.g.: gold
2. Solution C
3. Copper oxide
4. iron sulfate; copper

Page 160 – Key Words Exercise

Alkaline – A solution that has a pH greater than 7

Displaces – Takes the place of

Electrical conductor – A material that electricity can pass through easily

Molten – Liquid

Reactivity series – A list that places the metals in order from the most reactive to the least reactive

Rust – A compound formed when iron reacts with oxygen and water

Tarnish – Become dull

Thermal conductor – A material that heat can pass through easily

Vigorously – Quickly

Page 160 – Comprehension

1. A mixture of metals
2. They are made from different metals in different proportions.
3. a) It's very resistant to corrosion.
    b) To make cutlery and saucepans.
4. Hydrated iron oxide
5. It stops the oxygen and water from reaching the iron.

Page 161 – Testing Understanding

1. a) air / water; water / air; dull; rust; unreactive
    b) magnesium chloride
    c) oxygen; oxides; least
    d) more; less; more; iron sulfate; copper

2. a) Any one from: a thermometer; a data logger
    b) A measuring cylinder

c)

Metal Used	Temp. at Start (°C)	Temp. at End (°C)	Temp. Change (°C)
Magnesium	20	23	3
Copper	21	21	0
Zinc	21	23	2

d)
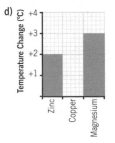
Type of Metal

Page 162 – Skills Practice

1. a) A thermometer
    b) Any one from: it logs the results; it's more accurate / precise; it measures more temperature points over time.

2.

Factor	Is it Changed?	Is it Measured?	Is it Kept the Same?
Temp. rise		✓	
Type of acid			✓
Type of metal	✓		
Volume of acid			✓

3.

4. Magnesium, zinc, iron, copper

ENVIRONMENTAL CHEMISTRY

Page 166 – Quick Test

1. A dark, sticky material, consisting of the remains of dead plants and animals, that holds pieces of rock together.
2. Add manure – as it rots down, acidic compounds are released.
3. Because carbon dioxide dissolves in it.
4. When sulfur oxides or nitrogen oxides dissolve in water.
5. Any sensible answer, e.g.: by lightning strikes
6. Any sensible answers, e.g.: methane; carbon dioxide

Page 167 – Key Words Exercise

Acid rain – A type of rain formed in polluted areas when sulfur oxides or nitrogen oxides dissolve in water

Carbonic acid – A chemical, formed when carbon dioxide dissolves in water, which makes normal rainwater slightly acidic

Deforestation – The cutting down of large numbers of trees

Fossil fuels – Non-renewable energy resources, such as coal, oil and natural gas, formed from plants and animals that lived long ago

Fungi – A type of microbe that decomposes plant and animal material

Greenhouse effect – How gases, including carbon dioxide and methane, cause the Earth to warm up

Humus – Dark, sticky material formed when plant and animal matter rots

Neutralisation – The reaction between an acid and an alkali

pH – A scale used to measure how acidic or alkaline something is

Quicklime – Calcium hydroxide

Satellite – An object that orbits a larger body

Sulfur dioxide – A toxic gas formed when sulfur is burned

Weathered – When a rock is broken down into smaller pieces

Page 167 – Comprehension

1. Spiders and insects

2. They're used to make some sunscreens and antibiotics.

3. A partnership where both partners benefit by the presence of the other.

4. The alga photosynthesises to provide food while the fungus provides a home for both plants.

5. That the level of air pollution is increasing.

Page 168 – Testing Understanding

1. a) pH; alkaline; acidic; quicklime; alkaline; lower
   b) carbon; acid; nitrogen; marble
   c) global; carbon; energy

2. a) Parts per million
   b)
   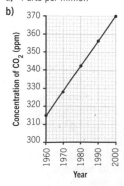
   c) It's increasing
   d) More fossil fuels are being burned; deforestation.

Page 169 – Skills Practice

1. a) and b)

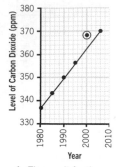

   c) The result for the year 2000 should be circled. It could have been misread; there could have been a problem with the equipment used; a lot of carbon dioxide was in the air that particular day.

2. **Any answer from 390ppm to 392ppm**

3. a) **Any one from:** it would be more reliable; any anomalous results will be easy to spot; the pattern would be easier to see.
   b) It would take more time / be more expensive to take more readings.

USING CHEMISTRY

Page 173 – Quick Test

1. A substance that can be burned to release heat energy.

2. Water (vapour) and carbon dioxide

3. Water (vapour)

4. In factories

5. In a reaction, the total mass of the reactants is equal to the total mass of the products.

6. It combines with oxygen, and oxygen has mass.

Page 174 – Key Words Exercise

Complete combustion – When a material is completely burned in a good supply of oxygen

Displacement reaction – A reaction in which a more reactive metal takes the place

of a less reactive metal

Fuel – A substance that can be burned to release heat energy

Hydrocarbon – A compound that contains only carbon and hydrogen

Incomplete combustion – When a material is burned in a limited supply of oxygen

Limewater – A solution of calcium hydroxide that turns cloudy if carbon dioxide gas is bubbled through it

Photosynthesis – The process by which green plants make glucose

Synthetic material – A material that's made in a factory

Page 174 – Comprehension

1. In 1733

2. His mother died when he was seven years old.

3. It's needed for things to burn and for iron to rust.

4. In the manufacture of fizzy drinks.

5. He expressed support for the French Revolution and his home was attacked by an angry mob.

6. He established a church and continued his scientific work to improve people's lives.

Page 175 – Testing Understanding

1. a) fuel; hydrocarbons
   b) incomplete; monoxide
   c) copper; magnesium sulfate; heat / thermal; temperature; energy / heat
   d) made / formed / produced; reactants; products; rearranged

2. a)
   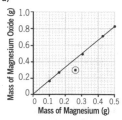

   b) The greater the mass of magnesium burned, the greater the mass of magnesium oxide produced.

   c) The result at 0.26, 0.32 should be circled. The results may have been misread; some of the magnesium oxide may have escaped.

Page 176 – Skills Practice

1. a) Mg
   b) Cu
   c) Fe

2.

Metals Used	Voltage (V)
Iron and copper	0.78
Magnesium and copper	2.71
Iron and magnesium	1.93

3.

Variable Changed	(Combination of) metals used
Variable Measured	Voltage
Variable Kept the Same	Type of fruit

4.
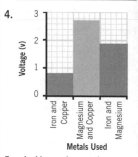

5. a) Magnesium and copper
   b) Iron and copper
   c) The greater the difference in reactivity, the greater the voltage produced; the smaller the difference in reactivity, the lower the voltage produced.

FINDING ENERGY

Page 181 – Quick Test

1. Change or transfer energy.

2. **Any one from:** Biofuel; wind energy; solar power; hydro-electric.

3. You will gain weight.

4. **Any one from:** Double glazing; cavity walls; cavity wall insulation; insulating the loft; draught excluders; carpets and curtains.

5. Efficiency = $\dfrac{\text{Useful energy}}{\text{Total energy used}}$

Page 182 – Key Words Exercise

Chemical energy – The energy stored in food and fuels
Efficiency – A way of measuring how much energy is wasted
Energy – This can't be created or destroyed, only changed
Fossil fuels – Oil, gas and coal are these, for example
Generator – It changes kinetic energy into electrical energy
Global warming – Heating of our planet as a result of polluting gases
Insulation – This reduces the amount of heat energy wasted
Kinetic energy – The type of energy possessed by moving objects
Renewable energy – Energy resources that can be easily replaced

Page 182 – Comprehension

1. Heat energy and electrical energy.

2. Because he was only an amateur scientist and his ideas were very different from what most scientists thought at the time.

3. His work fitted in with other ideas that other physicists were discovering.

4. Energy can not be created or destroyed, only changed.

Page 183 – Testing Understanding

1. a) gas; chemical; heat; electrical; heat
   b) roof; walls; cavity; roof insulation
   c) double-glazing; fuel; energy; pollution.

2. a) Fossil fuel
   b) Wind is free.
   c) When the weather is not windy other energy sources will have to be used.
   d) The setting-up / standby cost is lower; nuclear power is more reliable.

Page 184 – Skills Practice

1. Set up the apparatus as in the diagram. Light the substance and hold the test tube above the burning substance.

2. a) The hotter the water gets, the more energy is being released by each substance.
   b) The temperature increase of the water.

3. a) Use the same volume of water at the same initial temperature for each test. Make sure that the distance the test tube is held above the substance is kept the same in each test. Use the same mass of each substance to be tested. Use a new test tube for each test.
   b) Volume of the water, initial temperature of the water, mass of the substance, distance of the test tube from the substance.

4. Hold the test tube with tongs, point the test tube opening away from your body, place a heatproof mat below the substance, wear goggles, tie loose hair and clothing back. Follow usual laboratory safety procedures.

5. a) Heat / energy loss is the main source of error.
   b) Errors can be reduced by ensuring that the test is a fair test and making the variables in question 3 are kept the same.

6. In a table and / or a graph.

ELECTRICITY IN CIRCUITS

Page 188 – Quick Test

1.

2. Increase the voltage / number of cells; reduce the number of bulbs in the circuit.

3. True

4. A thin piece of wire that melts if the current is too high.

Page 189 – Key Words Exercise

Ammeter – This measures the current in a circuit
Amp – The unit of current
Appliance – Useful device that uses electricity
Battery – Two or more cells joined together
Bulb – This changes electrical energy into light and heat
Cell – This supplies energy for the electricity in the form of voltage
Circuit – This has to be complete for electricity to flow
Conductor – A material that allows electricity to flow through it
Current – The flow of electricity
Fuse – A safety device that melts when the current gets too high
Mains – High-energy electricity supplied to our homes
Parallel circuit – A circuit in which each bulb has its own loop
Series circuit – A circuit in which everything is connected together in one loop
Switch – Allows the circuit to be turned on and off.
Voltage – The electrical 'push' provided by a cell

Page 189 – Comprehension

1. It condenses into a liquid.

2. The amount a material opposes an electrical current.

3. It reduces to almost zero.

4. MRI scanners in hospitals and in Maglev trains.

Page 190 – Testing Understanding

1. a) series; both
   b) bulbs; go out; broken
   c) parallel; battery; switches
   d) lights; bulbs; quicker

2. a)
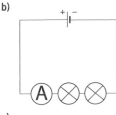
   b)
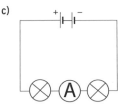
   c)
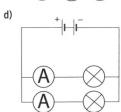
   d)
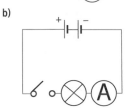

Page 191 – Skills Practice

1. a)

   b)
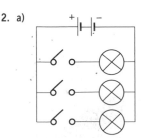

2. a)

b)

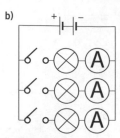

3. In parallel the current splits and some goes through each bulb. The total current is equal to the sum of the currents through each bulb. The current through each bulb is the same in each circuit. The total current in the circuit increases each time a bulb is added.

4. The cells will run down quicker with more bulbs in a parallel circuit because the total current drawn from the cells is greater.

5. The bulbs and the wires will get hot. This will change their resistances and affect the readings.

WHAT FORCES DO

Page 195 – Quick Test

1. Newton

2. Impact force

3. He falls at a constant speed (neither speeding up or slowing down).

4. How heavy something is in relation to its size.

Page 196 – Key Words Exercise

Balanced forces – When two forces acting on an object cancel out each other
Contact force – Forces that involve objects touching
Drag – Friction between a moving object and the air
Field – The area where a non-contact force acts
Impact – A contact force when objects collide
Mass – 1kg of this has a weight of 1ON
Stationary – Not moving

Tension – The force when a spring or a rubber band is stretched
Thrust – A force provided by an engine
Upthrust – The upward force from a liquid that can cause an object to float
Weight – The pull of the Earth on a mass due to gravity

Page 196 – Comprehension

1. The air.

2. A force.

3. A force.

4. It would keep on going forever.

5. Because on Earth objects stop moving due to friction, which you can't see.

Page 197 – Testing Understanding

1. **a)** friction; contact
   b) non-contact; field; magnetic; gravitational
   c) Newtons; weight
   d) Weight; 300N

2. **a)** **i)** Newtons
   ii) centimetres
   b)

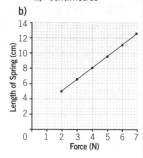

   c) As the force increases, the length increases; it is a linear relationship.
   d) 2cm

Page 198 – Skills Practice

1. It will avoid any water being lost by splashes and the plasticine will be easier to remove.

2.

3. Make sure the measuring cylinder is on a flat surface; read the measurement at eye level; read the measurement from the bottom of the meniscus (the curve the water makes at the surface).

4. Approx 1.7g/cm^3.

5. Less, because the plasticine sinks in the water.

BEYOND OUR PLANET

Page 202 – Quick Test

1. 1 day / 24 hours.

2. $365\frac{1}{4}$ days / one year

3. **Any two from:** sending and receiving television and telephone signals; taking pictures of weather patterns / weather forecasting; navigation; research.

4. Mercury

Page 203 – Key Words Exercise

Asteroid Belt – Area between Mars and Jupiter where lumps of rock are found orbiting the Sun
Axis – The line around which a planet rotates
Constellation – A pattern of stars that can be seen in the night sky
Galaxy – The Milky Way, for example
Geostationary orbit – An orbit in which the satellite stays in a fixed position relative to the Earth
Hemisphere – Half of the Earth, either the North or the South part
Moon – A natural satellite that orbits the Earth once every 28 days
Phase – Visible part of the Moon at night that changes during a month.
Planet – Object that orbits a star; the Earth, for example
Satellite – An object that orbits a planet
Solar System – The collection of our Sun and the planets
Stars – Balls of burning gas found in clusters called galaxies
Sun – A fiery ball of burning gas found at the centre of our Solar System

Page 203 – Comprehension

1. An object that gives out its own light.

2. Gravity.

3. Nuclear reactions.

4. A large star dying with an enormous explosion.

5. Very large stars.

Page 204 – Testing Understanding

1. **a)** hours; day
   b) orbit; year
   c) **i)** hemisphere; summer; winter
   ii) Southern; summer; winter
   d) Moon; axis; Earth

2. **a)** Planet A, it has a rotation time (a day) of 24 hours
   b) Planets C and D
   c) Planet B

Page 205 – Skills Practice

1. Distance from the Sun and surface temperature

2. a) and b)

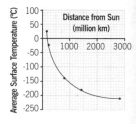

3. The average surface temperature of a planet decreases as its distance from the Sun increases.

4. **a)** Anything between 50°C and 150°C.
   b) The elements it is made of, or its atmosphere / lack of atmosphere.

GETTING HOTTER, GETTING COLDER

Page 209 – Quick Test

1. Degrees Celsius (°C).

2. The full bath at 25°C.

3. From hot to cold.

4. Convection; conduction; radiation.

5. Infrared waves.

6. Light colours reflect the radiation and help you to keep cool.

Page 210 – Key Words Exercise

Conduction – The way heat flows through a solid
Convection – The way heat travels through gases or liquids
Degrees Celsius – The unit of temperature
Global warming – The Earth getting hotter as a result of the greenhouse effect
Heat – A type of energy
Infrared – Heat radiation waves
Insulation – This reduces the amount of heat flow
Radiation – The way heat travels through air or a vacuum
Temperature – A measure of the hotness of a body
Thermometer – An instrument used to measure temperature

Page 210 – Comprehension

1. It transfers chemical energy in food to heat energy in the muscles.

2. About 85%.

3. 35°C

4. Wrapping them in lots of layers of clothing that trap air; wrapping them in a shiny blanket.

5. It shivers; the blood vessels shrink.

Page 211 – Testing Understanding

1. a) solids; conducted / transferred; good
   b) wooden; insulators
   c) liquids; convection; vacuum; radiation

2. a) Convection; conduction
   b) It reflects heat, preventing radiation.
   c) convection current.
   d) It's not a good insulator, it conducts heat well.
   e) It would keep a drink cool because it would prevent heat getting into the

flask by conduction and convection. To prevent heat getting into the flask by radiation, the outside should be silver rather than the inside.

Page 212 – Skills Practice

1. **Any sensible answers, e.g.:** The thickness of the material; the size of the beaker; the volume of water.

2. It traps the air, reducing convection.

3. It reflects the heat, reducing radiation.

4. Time (seconds); Temperature (°C)

5.

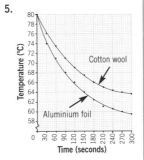

6. The cotton wool.

7. The water isn't cooling at a uniform rate in the beakers. If the water in the beakers was stirred more, then the curve should be smoother.

8. It would improve insulation because the gaps between the layers would trap air and reduce convection. The amount of energy transferred by conduction would be reduced because of the air gaps.

MAGNETISM AND ELECTROMAGNETISM

Page 216 – Quick Test

1. A magnetic field.

2. A tiny magnet that turns on a spindle. The north-seeking pole turns so that it points to the Earth's magnetic North.

3. Domains.

4. It creates two smaller magnets.

5. A coil of wire that becomes magnetic when a direct current is passed through it.

Page 217 – Key Words Exercise

Attract – A force that pulls together
Circuit breaker – A safety device that switches off with high current
Coil – If a current is passed through this it becomes an electromagnet
Core – This will make an electromagnet stronger
Domains – Tiny magnetic areas inside some metals
Electromagnet – A magnet that can be switched on and off
Field lines – These show the strength and direction of a magnetic field
Magnetic field – The area around a magnet
Repel – A force that pushes away
Soft iron – This can be magnetised but it doesn't retain its magnetism

Page 217 – Comprehension

1. It accelerates particles at very high speeds.

2. To smash them into other particles so that they break up in order to discover new particles and their behaviour.

3. It accelerates the particles round and round.

4. The Large Hadron Collider.

Page 218 – Testing Understanding

1. a) safety; electromagnet; high
   b) contact switch; current; attract; spring; circuit
   c) reset; flow

2. a)

   b) You can switch an electromagnet on and off.
   c) Wind some insulated wire around a soft iron core and pass direct

current through the wire.
   d) When the electromagnet attracts the soft iron, the contacts are pulled apart, breaking the circuit. The current stops flowing and the electromagnet loses its magnetism.
   e) **Any sensible answers, e.g.:** A circuit breaker; lifting cars in a scrap yard.

Page 219 – Skills Practice

1. Paper clips; an iron nail; insulated wire; battery (or other direct current supply).

2. Wind the wire around the nail four turns. Switch on the current and test the electromagnet to see how many paper clips can be suspended. Repeat the experiment for a range of number of turns, each time recording the number of paper clips that can be suspended.

3.

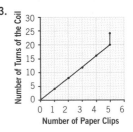

4. It is proportional up until 20 turns of the coil and then stays the same.

5. The wire might get hot and energy will be lost; the paper clips might not all be exactly the same.

6. Increase the current.

LIGHT RAYS

Page 223 – Quick Test

1. An object that gives off light.

2. About 8 minutes.

3. It's scattered.

4. It's the same size as the object; it's the same distance from the mirror as the object.

5. Because the light speeds up and is bent or refracted as it leaves the water, causing the water to look shallower than it really is.

Page 224 – Key Words Exercise

Incident ray – A ray that strikes a surface
Luminous – An object that gives out its own light
Normal – A line drawn at right angles to a surface at the point where the incident ray hits the surface
Opaque – A word describing a material that doesn't allow light to pass through it
Reflection – Light rays that bounce off a surface
Refraction – The bending of light when it changes speed
Scattering – Light reflected in all directions
Shadow – An area where there's no light
Translucent – A word describing a material that allows some light to pass through it
Transparent – A word describing a material that allows light to pass through it
Virtual – An image that can't be focussed on a screen

Page 224 – Comprehension

1. Because he used his ideas to explain reflection and refraction, and he was well known for other impressive scientific ideas.

2. Other scientists showed behaviour of light that could only be explained with wave ideas.

3. Einstein proved that light behaved as a particle, using evidence from an experiment carried out by Max Planck, a German physicist.

4. Some of light's behaviour can be described with waves, and some with particles.

Page 225 – Testing Understanding

1. a) luminous; reflects; eyes
   b) Transparent; opaque; translucent

c) straight; no; shadow

2. a) to d)

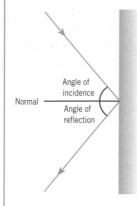

Normal
Angle of incidence
Angle of reflection

Page 226 – Skills Practice

1. A protractor.

2. They can take an average of the results, which will be more accurate than only one result.

3.
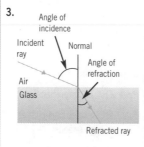

Angle of incidence
Incident ray
Normal
Angle of refraction
Air
Glass
Refracted ray

4. The glass beaker will also refract the light; the edge of the beaker is curved.

MAKING SOUND, HEARING SOUND

Page 230 – Quick Test

1. No, they only move to one side and then to the other side; they don't travel.

2. Energy.

3. It senses which ear is closer to the sound. It then turns its head to make another measurement.

4. There are no air particles to vibrate and pass on energy.

Page 231 – Key Words Exercise

Amplitude – The maximum distance from rest position
Auditory nerve – This sends electrical impulses to the brain
Cochlea – A coiled membrane full of liquid and nerve cells
Echolocation – Measuring distances by detecting echoes
Frequency – The number of complete waves per second
Hammer – A bone in the middle ear
Microphone – This converts sound energy into electricity
Oscilloscope – A sound wave can be viewed on one of these
Rest position – The normal position of a particle when it's not moving
Wavelength – The distance between a point on a wave and the same point on the next wave

Page 231 – Comprehension

1. A perforation.

2. The nerve endings can be damaged.

3. Near machines; airports; areas of loud music.

4. By wearing ear protectors.

Page 232 – Testing Understanding

1. a) canal; outer
   b) drum; cochlea; stirrup
   c) cells / endings; impulses
   d) direction; ear
   e) water; faster; time

2. a) The number of waves per second.
   b) Grasshopper; bat; dolphin.
   c)
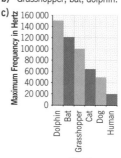

Maximum Frequency in Hertz
160 000
140 000
120 000
100 000
80 000
60 000
40 000
20 000
0
Dolphin
Bat
Grasshopper
Cat
Dog
Human
Organism

   d) **Any answer between 20 000 Hz and 50 000 Hz.**

Page 233 – Skills Practice

1. A microphone.

2. The maximum distance the wave reaches from rest position.

3. By measuring the maximum distance the waveform reaches from the rest position. This can be done accurately by measuring from the top to the bottom of the wave and dividing by two.

4. Plot a graph of both sets of results on the same axis and see if there's a pattern.

5. There seems to be a pattern between the two sets of results, they're approximately proportional.

ENERGY AND ELECTRICITY

Page 237 – Quick Test

1. **Any three from:** the wind; the Sun; water in rivers and reservoirs; tides in the sea

2. **Any three from:** in fuels; in batteries; in our bodies; in food

3. Volts

4. Direct current / current that flows in one direction.

5. Alternating current / current that changes direction.

Page 238 – Key Words Exercise

Alternating current – Current that continuously changes direction, created by a generator
Brushes – These ensure continuous contact of the wires in a generator as the coil rotates
Device – Something that transfers energy from one form to another
Direct current – Current supplied by a cell or battery
Double insulated – A device with a plastic casing
Generator – A coil of wire rotating between the poles of a magnet
Mains electricity – This electricity comes to your home at 230–240V
National Grid – A network of cables taking electricity from power stations to homes
Potential difference – This tells you how much energy is transferred in a cell or in a bulb in an electric circuit

Substations – The parts of the National Grid that contain step-up or step-down transformers
Transformer – This changes the voltage to a safer level for use in homes
Turbine – This usually drives a generator in a power station

Page 238 – Comprehension

1. It can increase the greenhouse effect and lead to global warming.

2. They change the landscape; they can disturb wildlife; they only produce a small amount of electricity.

3. a) **Any one from:** it releases no harmful gases into the atmosphere; it's a very efficient way of producing electricity.
   b) **Any one from:** waste material can be expensive to dispose of safely; the risk of accidental emission of radioactive material.

Page 239 – Testing Understanding

1. a) power stations; cables; National Grid; pylons; high
   b) electricity; homes; current; less; heat
   c) step-down; substations; safer; mains

2. a)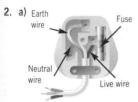
   Earth wire
   Fuse
   Neutral wire
   Live wire

   b) The live wire controls the AC current, flowing backwards and forwards many times per second. The neutral wire completes the circuit.
   c) Devices that are cased in plastic. Because plastic is an electrical insulator, this adds another layer of insulation to the device – hence the name double insulation.
   d) The earth wire

Page 240 – Skills Practice

1.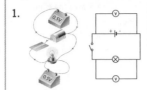

2. The amount of energy transferred in the bulb from the current to heat and light.

3. They should read the same value.

4. This shows that energy is conserved in the circuit.

5.

6. Both values will increase. If the second cell is identical to the first, both values will double.

PUSHING AND TURNING

Page 244 – Quick Test

1. Newtons per square metre (N/m^2) or newtons per square centimetre (N/cm^2)

2. The size of the force applied and the area that the force is acting on.

3. The size of the force and the distance from the force to the pivot.

4. 10Nm

5. 6Nm

Page 245 – Key Words Exercise

Atmospheric – The pressure of the air around us
Balanced – When the moments of the forces are equal
Collide – This is how gas particles exert pressure on the walls of a container
Dam – This is wider at the bottom because of the increase in the pressure of the water at a greater depth
Fulcrum – Another word for pivot
Hydraulic – A system that transmits pressure through a liquid

Lever – This uses the principle of moments to lift a heavy load
Moment – The turning effect of a force
Pivot – The point from which a turning force acts
Pressure – This is equal to force / area

Page 245 – Comprehension

1. Lever means 'to lift' in French and levers are often used to lift things.

2. **Any sensible answers, e.g.:** door handles; staplers; bike gears; car jacks

3. If a lever was long enough, he would be able to produce a large enough force.

4. The huge stone blocks are very heavy and would have been very difficult to move. Levers would have made them easier to move.

Page 246 – Testing Understanding

1. a) force; pressure; area
   b) increased; Studs; less; ground; knife
   c) increasing; reduce; sinking; Camels'

2. a) 1200Nm
   b) 800Nm
   c) The moments aren't the same.
   d) The boy should be moved so he is 3.0m from the pivot.

Page 247 – Skills Practice

1. The temperature of the gas would be much hotter at the bottom if heated directly. The water allows the heat to be distributed more evenly.

2. **Any sensible answer up to a maximum of 2 minutes,** e.g.: Every 30s

3.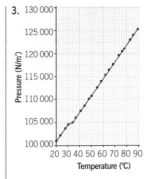

4. As the temperature of the gas increases, the pressure increases. They're proportional.

SPEEDING UP, SLOWING DOWN

Page 251 – Quick Test

1. Distance and time

2. 25m/s

3. **Any sensible answers, e.g.:** a ball slows down as friction acts on it; a ball stops as a goalkeeper catches it.

4. **Any sensible answers, e.g.:** a car accelerates as the engine drives it forwards; a planet changes direction as it orbits the Sun.

   (Note that the answers to questions 3 and 4 can be the same).

Page 252 – Key Words Exercise

Acceleration – A change in velocity
Balanced – When two forces in opposite directions are equal
Drag force – The force experienced by a moving object in a gas or liquid that opposes its motion
Friction – A force that opposes motion and produces heat energy
Gradient – Equal to velocity on a distance–time graph
Speed – The distance an object moves in a certain time
Stationary – An object that's not moving
Streamlined – An object that's shaped so that the drag force is less

Terminal velocity – The maximum velocity reached by a skydiver when the drag force balances the weight
Unbalanced – When two forces in opposite directions aren't equal
Unit – This must be the same when comparing the speeds of two different objects
Velocity – Speed in a specific direction

Page 252 – Comprehension

1. In the opposite direction to the motion.

2. It increases it.

3. It has a pointed nose and a streamlined torpedo shape.

4. Designers of submarines, boats and cruise liners can improve their designs and make them faster and more efficient.

Page 253 – Testing Understanding

1. a) time; distance
   b) direction; unit; second
   c) Acceleration; force; speed / direction; direction / speed

2. a) The velocity also starts off at almost zero and increases over the first 40s.
   b) They are unbalanced.
   c) i) 50m/s
      ii) It's constant at 50m/s.
   d) They are balanced.

Page 254 – Skills Practice

1. A stopwatch

2. **Any sensible answer, e.g.:**

Animal	Distance (m)	Time (s)	Time (s) (repeated)	Time (s) (average)	Speed of Plasticine Falling (m/s)
Cheetah					
Dolphin					
Elephant					
Eagle					

3. The idea is to test the streamlining of each shape, so it's fairer to keep the mass the same and change the shape. Otherwise, the test will be investigating a relationship between the size or mass of the animal and its speed.

4. The faster animals have more streamlined shapes.

5. **Any sensible answers, e.g.:** some animals travel in water, others in air, others on land; the animals are different sizes; the animals have different strengths.

SPACE AND GRAVITY

Page 258 – Quick Test

1. Copernicus

2. The church leaders, because they believed that God had placed the Earth at the centre of the Universe.

3. Their mass and the distance between them.

4. Because the apples only have a small mass.

Page 259 – Key Words Exercise

Big Bang – Scientists' model for the beginning of the Universe
Black hole – The gravity of this is so strong that not even light can escape
Geocentric – A model of the Solar System with the Earth at the centre
Heliocentric – A model of the Solar System with the Sun at the centre
Helium – The product of a nuclear reaction inside a burning star
Hydrogen – The fuel of a star
Nebula – A cloud of gas and dust
Neutron star – The dense core of a star left after a super nova
Red giant – A star that has run out of fuel becomes this
Super nova – A massive star dies in a big explosion known as this
White dwarf – The small, dense core of a dying star

Page 259 – Comprehension

1. He observed nebula in galaxies that were outside the Milky Way.

2. He proved that the Universe was much larger than people believed.

3. The Big Bang theory

4. It can be repaired by astronauts whilst in orbit.

Page 260 – Testing Understanding

1. a) gravity; temperature; nuclear; energy; helium
   b) core; hydrogen; red; white
   c) explode; gravity; light

2. a) It is strongest.
   b) It is weakest.
   c) It is fastest.
   d) It is slowest.

Page 261 – Skills Practice

1. The mass of the Plasticine

2. The force

3. a) She could ask a friend to help her spin the masses at the same speed, for example, once every second. As long as the length of the string is kept constant (and therefore the circumference of the circle), the speed will be constant.
   b) The length of the string

4. If the mass of Plasticine came off the string, would it hit anyone or anything breakable?

5. In a table and on a graph

Index

Index

Index

Index